WJEC
Physics
AS Level
Revision Workbook

Gareth Kelly

Iestyn Morris

Nigel Wood

Published in 2021 by Illuminate Publishing Limited, an imprint of Hodder Education, an Hachette UK Company, Carmelite House, 50 Victoria Embankment, London EC4Y 0DZ

Orders: please contact Hachette UK Distribution, Hely Hutchinson Centre, Milton Road, Didcot, Oxfordshire, OX11 7HH. Telephone: +44 (0)1235 827827. Email: education@hachette.co.uk. Lines are open from 9 a.m. to 5 p.m., Monday to Friday. You can also order through our website: www.hoddereducation.co.uk

British Library Cataloguing in Publication Data

A catalogue record for this book is available from the British Library

ISBN 978-1-912820-62-7

Printed by Ashford Colour Press Ltd

Impression 3
2023

Hachette UK's policy is to use papers that are natural, renewable and recyclable products and made from wood grown in well-managed forests and other controlled sources. The logging and manufacturing processes are expected to conform to the environmental regulations of the country of origin.

Publisher: Eve Thould

Editor: Geoff Tuttle

Design and layout: Nigel Harriss

Cover image: Shutterstock.com/Richard A McMillin

Acknowledgements

The authors would like to thank Adrian Moss for guiding us through the early stages of production and the Illuminate team of Eve Thould, Geoff Tuttle and Nigel Harriss for their patience and meticulous attention to detail. We are also indebted to Helen Payne for her eagle eyes in spotting mistakes and inconsistencies and for her many insightful suggestions.

Picture Credits

p122: Courtesy of School of Physics and Astronomy, University of Minnesota.
Other illustrations © Illuminate Publishing

Contents

How to use this book

What this book contains

The AS physics course, which is also the first year of the A level physics course, consists of two units, each assessed with a 90-minute exam.

Unit 1: Motion, Energy and Matter – divided into seven topics

Unit 2: Electricity and Light – divided into eight topics

Each topic has its own section in this book and there you will find:

- A concept map, which displays how the different concepts within the topic are related to one another and to other topics.
- A set of graded questions, with space for your answers, which are designed to test the content of the topic in a way that is similar to the examination.
- A question and answer section containing one or two examples of exam-style questions, with answers from two students, Rhodri and Ffion (who produce answers of different standard), together with examiner marks and discussion.

The next section comprises two practice papers – one for each unit. The final section consists of model answers to all the graded questions and the practice papers.

How to use this book effectively

This book can be used exclusively for revision, in which case you will work your way gradually through the questions as you revise each topic. Alternatively, it can be used regularly as you are working your way through AS physics, and each topic of questions can be used as an end of topic check/test. Your teacher might even like to use it as a homework book containing 15 sets of end of topic homework questions as well as the practice papers.

Probably the worst method of revising physics is simply reading your notes or your textbooks, a technique that will just send you to sleep and will not help you to retain more than a small fraction of the information. Regardless of the quality of writing of the notes/textbooks, there is an inherent soporific effect to reading notes. One way of combating this tendency to lose concentration is to make your own notes as you read through the notes/textbook. However, if you only reread notes, you will just prepare for exam questions in which you reproduce learned material. These so-called Assessment Objective 1 (AO1) questions form only 35% of the exams. The higher-level skills required for AO2 and AO3 require different revision techniques. See pages 5 to 8 for further information.

You will find that the best way of revising is to answer exam-type questions. As in many fields, it is practice that makes perfect in physics and you should practise as many past papers and questions within this book as you can.

You will, inevitably, come across questions that seem difficult. Should you not be able to answer them, it is time to visit your notes/textbook, but you will be doing so with a purpose to help you to concentrate. Should the question still leave you stumped, have a look at the answers in the back of the book. If the answer seems unintelligible, then it is time to ask your teacher or your fellow students for an explanation of how and why the answer is correct. Your teacher may also point out that different answers are creditworthy, especially in discussion questions.

So answering these questions and analysing the model answers will do more for your exam preparation than just reading notes or making revision timetables (however colourful)!

Assessment objectives

You need to demonstrate expertise in answering examination questions in different ways. One of the ways in which an exam question can be looked at is whether it is mathematical or not – in physics, there is maths in most questions. Some questions require expertise in practical skills, even in the written papers. However, besides these two categories there are the assessment objectives. There are three of them:

Assessment Objective 1 (AO1)

AO1 questions are ones in which you need to:

demonstrate knowledge and understanding of scientific ideas, processes, techniques and procedures

These questions account for 35% of the marks in the two AS unit papers. In the full A level the tally is 30% because the A2 units consist of more of the other assessment objectives.

The sentence in bold sounds far more complicated and highfalutin than is necessary. Essentially, these are the marks that can be obtained without too much thinking. This category covers:

- Recall of definitions, laws and explanations from the specification
- Inserting appropriate data into equations
- Deriving equations where required in the specification
- Describing experiments from the **specified practicals**.

In general, no judgement or new thinking is required. One can learn a definition or a law without fully understanding it, so it is a good idea to use the WJEC terms and definitions (T&D) booklet to help you memorise these. Defined T&D from the booklet are printed in **bold** in the concept map for each topic in this book.

Examples of AO1 questions

1. Explain the difference between a baryon and a meson. [2]

   Good answer (full marks 2/2): A baryon has 3 quarks whereas a meson has a quark and anti-quark.

   Bad answer (no marks 0/2): Baryons and mesons are made of quarks but one has 2 and the other 3.

   The *bad answer* misses the anti-quark point and it is not clear which of the two has 3 quarks.

2. State what is meant by the work function of a metal. [1]

   Good answer (1/1): The minimum energy required to remove an electron from a clean surface.

   Bad answer (0/1): The energy it takes to remove an electron from a metal.

   The bad answer misses the idea that it is the *minimum* energy and that the emission is from the surface.

3. Calculate the photon energy of e-m radiation of wavelength 3.0×10^{-10} m. [2]

   Good answer (2/2): $E = hf = \dfrac{hc}{\lambda} = \dfrac{6.63 \times 10^{-34}\ \text{J s} \times 3.00 \times 10^{8}\ \text{m s}^{-1}}{3.0 \times 10^{-10}\ \text{m}} = 6.63 \times 10^{-16}\ \text{J}$

   Bad answer (0/2): $E = hf = 6.63 \times 10^{-34} \times 3.0 \times 10^{-10} = 1.989 \times 10^{-43}\ \text{J}$

   Although the *bad answer* quoted a correct equation ($E = hf$), the student didn't insert the data correctly (the wavelength was inserted instead of the frequency) and so gained no credit. Putting the units in the working might have made it easier for the student to spot the mistake.

Assessment Objective 2 (AO2)

AO2 questions are ones in which you need to:

apply knowledge and understanding of scientific ideas, processes, techniques and procedures:

1. **in a theoretical context**
2. **in a practical context**
3. **when handling qualitative data**
4. **when handling quantitative data.**

Here, the key words are 'apply knowledge'. The application of knowledge is required in theoretical, practical, qualitative and quantitative contexts. Theoretical just means some idealised context made up by the examiner. Practical means that the data have apparently come from a real experiment (although the data are usually made up by the examiner). Qualitative means without numbers and calculations, whereas quantitative means the opposite (i.e. with numbers and calculations).

Note that application of knowledge here can also include analysis of data even though 'analyse' appears in AO3 (see page 8). Generally, if you are told what type of analysis to carry out, these will be AO2 skills. If the question is more open-ended and you must choose the analysis methods yourself, the question will be classified as AO3. AO2 is the most common type of question and accounts for 45% of the marks on the papers. Note that all calculations must be mainly AO2 marks: we have seen that inserting data into an equation is classed as AO1, but any manipulation of an equation, such as changing the subject, and the production of a final answer is AO2.

Examples of AO2 questions

1. A force of 6.96 N places a wire of cross-section 4.3×10^{-5} cm² under a strain of 2.6%. Calculate the Young modulus of the material of the wire. [3]

Good answer (3/3): $E = \dfrac{\sigma}{\varepsilon} = \dfrac{6.96\,\text{N}\,/_{4.3 \times 10^{-9}\,\text{m}^2}}{0.026} = 6.2 \times 10^{10}\,\text{Pa}$

Bad answer (1/3): $E = \dfrac{\sigma}{\varepsilon} = \dfrac{6.96\,\text{N}\,/_{4.3 \times 10^{-5}}}{2.6} = 6.2 \times 10^{4}\,\text{Pa}$

The two mistakes here are not to convert the cm² to m² and not expressing the 2.6% as 0.026.

2. A diffraction grating has 320 lines per mm. Calculate the angle of the 3rd order bright dot when light of wavelength 633 nm is shone upon it normally. [3]

Good answer (3/3): $d = \dfrac{1}{320\,000} = 3.125 \times 10^{-6}$; $\theta = \sin^{-1}\left(\dfrac{3 \times 633 \times 10^{-9}}{3.125 \times 10^{-6}}\right) = 37.4°$

Bad answer (0/3): $\theta = \sin^{-1}\left(\dfrac{3 \times 633}{320}\right) =$?? my calculator says it's an error.

The candidate does not understand the equation $\sin \theta = \dfrac{n\lambda}{d}$ because d is wrong and the wavelength is not converted from nm to m, so has not 'applied knowledge' correctly.

Assessment Objective 3 (AO3)

AO3 questions are ones in which you need to:

analyse, interpret and evaluate scientific information, ideas and evidence, including in relation to issues, to:
1. **make judgements and reach conclusions**
2. **develop and refine practical design and procedures.**

These questions account for 20% of the marks in the AS unit papers, 25% in the full A level.

The verbs *analyse*, *interpret* and *evaluate* are all appropriate and this is, indeed, what you will have to do. Most of these AO3 marks will concentrate on the first point – *judgements* and *conclusions*. The context will often be similar to one of the specified practicals with realistic data. Your analysis may well include analysing graphs to make numerical conclusions. You might have to evaluate the quality of the data and your conclusions. In some questions you are given a statement and have to determine whether or not (or to what extent) it is true. There are usually several ways of getting a sensible answer: you must choose one and structure your answer carefully. Other questions relate to the second part of the AO3 statement – develop and refine practical design and procedures. Usually, these questions are based on imperfections in the data and how you could improve the procedure or the apparatus to obtain better data. To answer these questions, you will need to read them carefully because there will be a clue (perhaps right at the start) as to what went wrong.

Another type of question is based on the part of the statement *'including in relation to issues'*. The **'issues'** include risks and benefits; ethical issues; how new knowledge is validated; how science informs decision making. Try and make sensible comments; the mark scheme will allow for many approaches and the marks will be quite attainable – approach the questions like a politician: have a view. Every theory paper has one issues question.

Examples of AO3 questions

1. Discuss the quality of data obtained by Gwynfor in order to calculate the Young modulus of copper and whether, or not, the data are in good agreement with theory. [4]

Good answer (4/4): The data appear good because all points lie close to the line of best fit. The best fit line starts off as a straight line through the origin, in agreement with theory, and the gradient decreases as measurements are taken past the elastic limit of copper, also in agreement with theory. The final value for the Young modulus of copper is 25% too low and is inaccurate (since the answer to the previous part suggested an uncertainty of only 15%).

Bad answer (0/4): I like the data because there is a definite pattern to the results and the graph is the same shape as the one that appears in the textbook. The final value of the copper YM seems too low and therefore is inaccurate.

2. Determine whether the light ray shown will propagate a long distance along the optical fibre. [5]

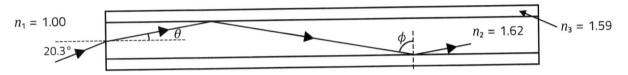

Discussion: This is an AO3 question because you are not told what principle to apply. To achieve good marks you must realise that total internal reflection (TIR) is involved, do the necessary calculations to determine whether the light ray undergoes TIR and explain how the results of the calculations inform the question.

Preparing for the examinations

Examination mark schemes

When examiners write questions for the AS examination, they also provides mark schemes containing details of how they are to be marked. For an example of a question and its mark scheme, see page 20. You'll notice that each part of the question is covered, with details of the sort of answer required for each mark. The mark scheme also contains information about the Assessment Objectives and any marks which count towards the mathematical and practical skills on the paper.

Let's look at this mark scheme in detail:

Part (a) is a piece of bookwork which you will be expected to know. The mark scheme gives an expected answer, but notice the phrase or *equivalent*. This means that the examiner (i.e. the marker) will look for a correct expression of this idea, however worded. For example, you might say, 'The sum of the components of the forces in any direction is zero'. All the markers are current or retired physics teachers, so they will know how to interpret a slightly different answer.

Notice also the word *vector*. It is in square brackets, meaning that the examiner would like to see it but, if you left it out, you would still get the mark.

The second and third marks are a bit more complicated: one mark is for the ideas which are not *underlined* and the other is for the italic words. Even when it doesn't say so, 'or equivalent' is always implied.

Part (b) is split into three parts. Notice in part (i)(I), you only get the mark if you give the correct unit.

In part (i)(II) notice the letters **ecf**. These stand for *error carried forward*. You need to know the value of two forces (11.8 N and 35.3 N) and you have just calculated them in part (b)(i)(I). If one or more of the answers in (b)(i)(I) was incorrect then this expression is a life-saver. The examiner will allow you to work with these in answering (b)(i)(II) and gain full marks, as long as you don't make any more mistakes!

The other thing to notice is the expression **or by implication**. This means that the examiner will award the mark even if you haven't written the equation down as long as your answer was correct – because it will be assumed that you must have applied correct physics in order to get the answer. Again, even when the mark scheme doesn't say so, the 'or by implication' rule applies unless you have been asked to **show that** or been told to **show your working**.

The marking

Now have a look at Rhodri's and Ffion's marked answers to this question. Notice that the examiner has put in ticks and crosses, where the mark has been given or withheld. You'll see also some annotations by the examiner. If you get a mark by ecf, the examiner will write this – see Rhodri's answer to (b)(i)(II)

On Rhodri's first answer, the examiner has written 'not enough'. This shows that the answer is almost there but there's too much missing for the mark to be given.

Another common annotation is bod – see Ffion's answer to part (a). This stands for *benefit of the doubt*. Ffion has missed out an important word (components) but the examiner considered that she had done enough to be credited.

Each of the AS unit papers is 1½ hours long and marked out of 80 marks.

Unit 1

Containing seven topics, it will have around 11 marks per topic, and you might expect the Unit 1 examination to contain seven questions – one on each topic. While every year's paper will differ significantly from this basic structure, the examiners will try to distribute their marks equitably between the topics, and seven questions will be a reasonable rule of thumb. However, there are four things (other than randomness of distribution) that arise to mess up this beautifully symmetric system.

1. Practical content: You can expect 20% of the examination (16 marks) to be based on experimental analysis. This usually means that one (or possibly two) of the questions will be based on one of the six specified practicals for this unit. This could be a description of the method, error analysis, graphs and conclusions – often the longest question on the paper.

2. Quality of extended response (QER): This is a 6-mark question with a lot of lines for writing and maybe some space for diagrams, too. These tend to be AO1 marks and so rely on you learning the basic physics required to answer the question. This, however, is only part of the problem. Not only must you put the required information down on paper but you must also do so in a logical, well-presented format employing good language skills. The penalty for poor spelling, punctuation and grammar is generally only 1 mark at most but the penalty for not knowing the relevant physics is 6 marks! A common type of question for this 6-mark QER is a description of one of the specified practicals.

3. Synoptic content: Although the Unit 1 examination is usually a few days before the Unit 2 examination, you still need to ensure that you have revised Unit 2 thoroughly because of this synoptic content. Any of the topics of Unit 2 can be combined with a Unit 1 topic to make a more difficult question, e.g. a question about energy might combine work, kinetic energy and electrical energy.

4. Issues: There is always going to be a question about 'issues' (including such things as decisions involving the use of physics in everyday life), and this will be 2 or 3 AO3 marks. You cannot revise for these questions but do practise the previous questions that have arisen. Just be confident and try to put down some sensible points leading to a sensible conclusion.

Unit 2

This unit contains eight topics leading to a mean of 10 marks per topic, and the rule of thumb this time will be eight questions. Note that, sometimes, two topics might be combined into one longer question, or one topic might be split into two smaller questions. Everything about practical content, QER, synopticity and issues applies equally to Unit 2, but there is one thing that can be added about the practical content.

Practical content in Unit 2: Examiners will make every effort to ensure that the practical skills examined in Unit 2 are different from those in Unit 1, e.g. plotting points will not be asked for in both units. The same goes for the other practical skills, such as measuring gradients, describing lines of best fit. So, after Unit 1 you will know what to look for in Unit 2.

Key command words and phrases in examination questions

These are the words or phrases which let you know what sort of answer is expected – there are quite a few to look out for.

State: Just provide a statement without an explanation.

> Example: State what happens to the current as the temperature of the metal wire increases.
>
> Answer: It decreases.

Define: You need to provide a statement which is close to (or equivalent to) that which appears in the WJEC Terms and Definitions booklet.

> Example: Define the potential difference (pd) between two points.
>
> Answer: It's the energy converted from electrical potential energy (to other forms) per unit charge (passing between the points).

Explain what is meant by (or explain the meaning of …): This can mean a couple of things:

> 1. Sometimes it just means the same as 'define',
>
>   Example: Explain what is meant by the potential difference (pd) between two points.
>
>   Answer: [Exactly the same as above.]
>
> 2. Sometimes it's a definition with a number included.
>
>   Example: Explain what is meant by the statement 'The Young modulus of steel is 2×10^{11} Pa'.
>
>   Answer: This is the stress divided by the strain **and** for steel it is 2×10^{11} Pa.

Explain the difference (between two things): This is two definitions in disguise because if you define both things you have automatically explained the difference between them.

> Example: Explain the difference between a vector and a scalar.
>
> Answer: A vector has magnitude and direction whereas a scalar only has magnitude.

Describe: Provide a brief description but no explanation is required.

> Example: Describe the appearance of an emission spectrum.
>
> Answer: Bright lines on a dark background.

Explain … (some statement): Sometimes this requires a logical argument.

> Example: Explain why a diffraction grating provides a more accurate value of the wavelength of laser light than a double slit even though the separation of the slits is known exactly for both.
>
> Answer: The diffraction grating provides brighter and sharper dots and hence, their position can be determined slightly more precisely. The diffraction grating will give a far greater separation of the dots meaning that the angle can be measured with a lower % uncertainty.

Suggest … (or suggest a reason …): Although not a common command word, this can produce some questions that are difficult to answer, often because candidates are not sure what is being asked. There will be more than one acceptable answer, otherwise the examiner would have used a different command word. These will often be AO3 marks, appearing at the end of a question requiring evaluation skills.

> Example: Suggest a reason why the gradient of the temperature against time curve decreases at higher temperature.
>
> Answer: The gradient decreases because more heat is being lost as the temperature difference between the aluminium block and the air increases (hence, more energy input is required for each unit temperature). An alternative answer would be that the specific heat capacity increases with temperature.

Calculate or determine: The aim is to obtain the correct answer (along with the correct unit if required by the mark scheme). With this command word, the correct answer will obtain full marks without the workings. However, you are advised strongly to show your working as marks are available for this even if the answer is wrong.

> Example: Calculate the mass of a 2.00 cm diameter steel ball of density 7800 kg m^{-3}.
>
> Answer: $m = \rho V = \rho \frac{4}{3}\pi r^3 = 7800$ kg m$^{-3} \times \frac{4}{3} \times \pi \times (0.0100 \text{ m})^3 = 0.033$ kg (2 sf)
>
> [Note that you do not have to put units in the calculation – but you do in your answer!]

Compare: Not a common command word but you ought to do what it says on the tin – compare the things it says to compare in the question.

> Example: Compare the appearance of a hot star (10 000 K) with that of a cool star (3000 K) with the same diameter.
>
> Answer: The hot star will appear brighter and slightly blue, and the cool star will appear red.

Evaluate: You will be required to make a judgement, e.g. whether a statement is correct or wrong, or to decide whether data are good or a final value is accurate.

Justify: This is sometimes used in a very similar manner to the word 'determine' when AO3 marks are being examined, e.g. justify Blodeuwedd's statement that the 2.00 V reading was anomalous.

Discuss: This can often be a command word in the 'issues' question. In general, you will not go far wrong if you make a couple of points in favour of the discussion issue, a couple against it, and then draw some sort of a sensible conclusion.

Common exam mistakes

1. **Not converting the given numbers correctly:** Planetary distances are usually in km while wire radii are in mm. Resistors can be in Ω, kΩ, or MΩ. These values have to be converted to the correct powers of ten. There are other common conversions such as changing diameter to radius when using area or volume formulae. All these are simple mistakes and do not show a poor understanding of physics. They are not penalised more than one mark most of the time. Nonetheless, these are probably the most common mistakes committed by physics students.

2. **Not reading the question carefully enough:** This usually results in not answering the question that was asked – either by answering a different question altogether or by missing part of the question. The most common parts of questions that are omitted are those that do not have dotted lines for you to answer on, e.g. adding to diagrams. Pay particular attention to these short parts of questions. Other common missed questions are ones that have an **and** condition in the question itself, e.g. calculate the magnitude **and** the direction. One or other part of the question will have been forgotten in the answer.

3. **Not understanding equations properly:** This often involves substituting wrong values into equations – an unforgivable sin! In kinematics equations, for instance, u and v are often mixed up. You shouldn't really have to use the data booklet; you should know the equations intimately and only check from time to time to ensure that you recollect them correctly. How do you ensure that you don't misunderstand an equation? Practise, practise, practise!

4. **Not knowing the basic terms and definitions** (a surprisingly common cause of loss of marks.) There is a WJEC booklet full of these – you should know everything within its covers.

5. **Forgetting to square a value in the equation:** This happens most often with the kinetic energy equation – the equation $E = \frac{1}{2}mv^2$ is written correctly but then the candidate forgets to square the velocity on the calculator. Or the converse: forgetting to square root the answer when using the same equation to calculate the velocity!

6. **Not planning the structure before answering the QER** (and extended explanations): Too many QER responses are rambling and unstructured. This is easily remedied by spending a moment to plan and structure your answer. Using short sentences tends to help, too.

7. **Not matching the correct corresponding values in a calculation:** By far the most common occurrence of this mistake is with electrical circuits: current, pd and resistance, e.g. a pd and a current will be combined to obtain a resistance ($R = V / I$) but the current and pd do not match – the pd is for one resistor and the current for another.

Unit 1: Motion, Energy and Matter

Section 1 Basic physics

Topic summary

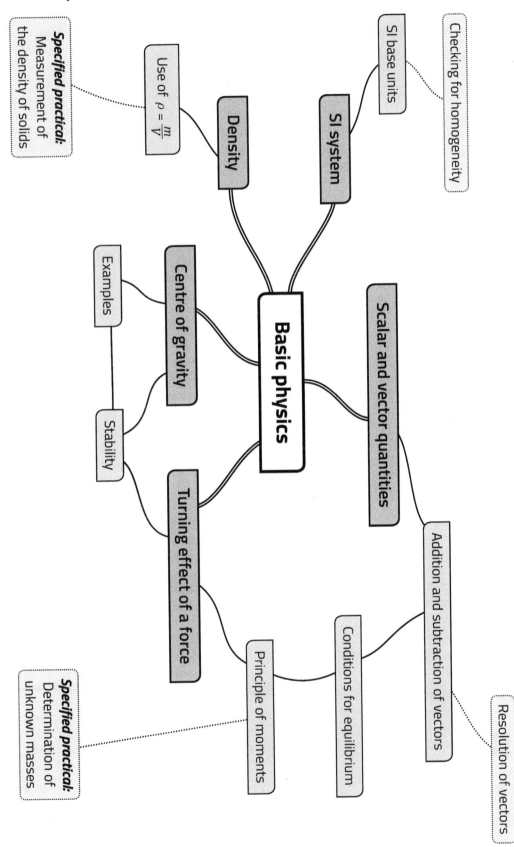

Specified practical: Measurement of the density of solids

Use of $\rho = \frac{m}{V}$

Density

SI base units

Checking for homogeneity

SI system

Basic physics

Scalar and vector quantities

Examples

Centre of gravity

Stability

Turning effect of a force

Addition and subtraction of vectors

Resolution of vectors

Conditions for equilibrium

Principle of moments

Specified practical: Determination of unknown masses

Q1 The candela (cd) is one of the seven SI base units. It is not used in A level physics. State the names and symbols of the other six SI base units. [2]

....

....

....

Q2 Newton's second law of motion can be written $F = ma$.

(a) State the SI unit of force and give its symbol. [1]

....

(b) Use the equation to write the unit of force in terms of the metre (m), kilogram (kg) and second (s). [2]

....

....

....

(c) Use the answer to (b) and the defining equation for work to show that the joule, J, can be expressed as $kg\ m^2\ s^{-2}$. [2]

....

....

....

Q3 Very slow-moving objects in air experience a drag force, F, proportional to their speed, v, through the air, that is $F = kv$, where k is a constant.

(a) Show that the unit of k can be written $[k] = kg\ s^{-1}$. [2]

....

....

....

(b) For more rapidly moving objects, the drag equation becomes $F = KAv^2$, where A is the cross-sectional area and K is a constant (different from that in part (a)). Express the unit of K in terms of the base SI units. [2]

....

....

....

Q4 The area, A, of a circle is related to its radius by the equation $A = \pi r^2$. Explain why π has no unit. [1]

....

....

Q5 The following is a list of quantities met in A level physics. Divide them into *scalar* and *vector* quantities.

[2]

energy   acceleration   time   density   temperature   velocity   momentum   pressure

....

Q6 One of the kinematic equations for constant acceleration is $v = u + at$. Show that this equation is homogeneous.

[2]

....

....

....

....

Q7 Two forces act on an object. The forces have magnitudes of 28 N and 45 N. Draw diagrams to show how the resultant force can have magnitudes of: (a) 73 N, (b) 17 N and (c) 53 N.

[4]

In each case, state the direction of the resultant force relative to the 45 N force.

[Hint: $28^2 + 45^2 = 53^2$]

(a)

Direction = ...

(b)

Direction = ...

(c)

Direction = ...

Q8 A force acts at 25° to the horizontal. The magnitude of the vertical component of the force is 53 N. Calculate:

(a) The magnitude of the force.

[2]

....

....

....

(b) Its horizontal component.

[1]

....

....

Q9 Two tugs, T_1 and T_2, are towing a large ship out of port. T_1 exerts a force of 5.0 kN at 15° to the forward direction. T_2 is a more powerful tug and exerts a larger force F at 10° to the forward direction. The resultant force of these is exactly in the forward direction.

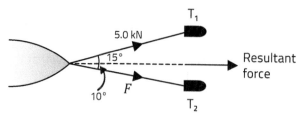

(a) Show that the magnitude of the component of the 5.0 kN force at 90° to the forward direction is approximately 1.3 kN. [2]

....

....

....

(b) State the magnitude of the component of force F at 90° to the forward direction. [1]

....

(c) Calculate the magnitude of force F. [2]

....

....

....

(d) Calculate the resultant of the two forces on the ship. [3]

....

....

....

....

Q10 An object has several forces acting on it. State the two conditions which must be satisfied for the object to be in equilibrium. [2]

....

....

....

....

....

Q11 A ball is thrown through the air. At time t_1 it is travelling with a velocity of 15.0 m s^{-1} at 30.0° to the horizontal and climbing. At time t_2 its velocity is 20.0 m s^{-1} at 49.5° to the horizontal and falling.

(a) Show that its horizontal and vertical components of velocity at t_1 are 13.0 m s^{-1} and 7.50 m s^{-1} upwards, respectively. [2]

....

....

....

(b) By considering the horizontal and vertical components of velocity at t_2 and your answers to part (a), calculate the change in velocity of the ball between t_1 and t_2. [3]

....

....

....

....

....

Q12 A ball, travelling at 12 m s^{-1}, collides with a wall and rebounds at 10 m s^{-1} at right angles to its original direction. The diagram shows this seen from above.

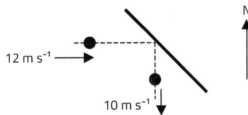

Calculate the magnitude and direction of the change in velocity. [3]

....

....

....

....

....

Q13 A spring has a spring constant k. This is the force per unit extension needed to stretch the spring and has the unit N m^{-1}. If an object of mass m is hung from the spring, pulled downwards and released, it oscillates with a period, T. A student cannot remember whether the correct equation for calculating T is:

$$T = 4\pi^2 \frac{m}{k}, \qquad T = 4\pi^2 \frac{k}{m}, \qquad T^2 = 4\pi^2 \frac{m}{k}, \qquad \text{or} \qquad T^2 = 4\pi^2 \frac{k}{m}$$

By considering the units of T, m and k, evaluate whether any of these equations could be correct. [3]

....

....

....

....

....

Q14 A reel contains copper wire with a diameter of 0.317 mm. Calculate:

(a) The cross-sectional area of the wire in m^2. Give your answer in standard form. [2]

(b) The volume of a 85.0 cm length of the wire. [2]

(c) The length of this wire that could be produced from a block of copper of mass 2.50 kg.
[The density of copper is $8.96 \times 10^3 \, kg \, m^{-3}$.] [3]

Q15 (a) A gas at a pressure of 2.50 MPa is held by a piston in a cylinder of radius 10.0 cm. Calculate the force that the gas exerts on the piston. [2]

(b) A sample of gas is held in an open-ended pipe by a piston consisting of a solid disc of copper, of length 10.0 cm, as shown. The pressure, p_A, of the air above the copper disc is 101 kPa. Calculate the pressure, p, of the gas. [Hint: the disc is held in position by a net upward force due to the pressure difference.] [4]

(Density of copper = $8.96 \times 10^3 \, kg \, m^{-3}$)

Q16 Rhian is given a rectangular block of iron and asked to determine its density. These are her measurements:

length / cm	6.35, 6.38, 6.34, 6.38, 6.37
width / cm	4.26, 4.24, 4.28, 4.17, 4.25
height / cm	2.79, 2.81, 2.83, 2.80, 2.81
mass / g	599.5

(a) Rhian thinks that she made a mistake with one of her measurements, so decides to ignore it. Identify the suspect measurement and give a reason. [2]

....

....

....

(b) In calculating the uncertainty in her value for the density of the iron, Rhian decides to ignore the uncertainty in the mass measurement. Evaluate her decision. [2]

....

....

....

(c) Use the data to calculate a value for the density of iron together with its **absolute** uncertainty. [5]

....

....

....

....

....

....

....

Q17 Maurice is given a uniform **half-metre** rule, a 100 g mass and a piece of cord and asked to measure the mass, m, of a piece of metal. He suspends the 100 g mass from the 1.0 cm mark and balances the ruler on the edge of a pencil. The balance point is 15.0 cm.

pencil

He replaces the 100 g mass with the piece of metal. The balance point is now 12.5 cm.

Use these results to determine the mass of the piece of metal. [4]

....

....

....

....

....

Q18 A uniform public house sign of width 80 cm and mass 3.5 kg is suspended from a metal bar of mass 1.5 kg and length 90 cm by two links, **A** and **B**, each of which is 10 cm from the edge of the sign. The metal bar is attached to the wall by a hinge, **H**. Link **A** is 15 cm from **H**. The metal bar is also attached to the wall by a wire which is attached to the bar at **B**.

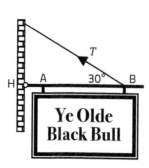

(a) Explain why the tensions in the two links, **A** and **B**, are equal and calculate their magnitude. [3]

(b) Calculate the sum of the clockwise moments, about **H**, of the forces on the bar. [2]

(c) By considering moments about **H**, calculate the tension, T, in the wire, stating which principle for equilibrium you use. [3]

(d) The hinge, **H**, also exerts a force on the metal bar. Using a different condition for equilibrium, determine the magnitude and direction of the force exerted by the hinge on the bar. [3]

Q19 Dominic completes a table of vector and scalar values:

Vectors	Scalars
Force	Temperature
Work	Displacement
Velocity	Energy

(a) State which of Dominic's choices are incorrect. [2]

....

....

....

(b) In fluid dynamics, the viscosity (μ) of a fluid can be given by the equation:

$$\mu = \frac{\rho u L}{k}$$

where k is a dimensionless number, ρ the density of the fluid, u the speed of the flow and L is a characteristic length.

(c) (i) Explain why $N\,s\,m^{-2}$ is a valid unit for the viscosity, μ. [4]

....

....

....

....

....

(ii) The dimensionless number, k, in the equation is an important factor in determining the motion of a cricket ball. Calculate k for a cricket ball with characteristic length (L) of 7.1 cm, travelling through air of density 1.16 $kg\,m^{-3}$ and viscosity $1.87 \times 10^{-5}\,N\,s\,m^{-2}$ at a speed of 41.2 $m\,s^{-1}$. [3]

....

....

....

....

....

Question and mock answer analysis

Q&A 1

(a) An object is acted on by forces whose lines of action are all in one plane. State the conditions necessary for the object to be in equilibrium. [3]

(b) In a theatre a spotlight of mass 3.60 kg hangs from a uniform rod of length 3.00 m and mass 1.20 kg. The rod is kept horizontal by vertical wires attached to points A and B as shown.

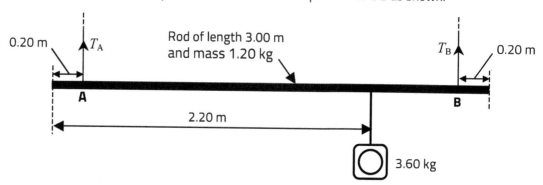

(i) (I) Use labelled arrows to show the magnitude and direction of the two other main forces on the system. [2]

(II) Determine the tensions T_A and T_B. [4]

(ii) Explaining your reasoning, determine the least mass that can be hung from *anywhere* on the bar to make the bar tilt. The 3.60 kg spotlight has been removed. [3]

What is being asked

This question is about the equilibrium of objects when acted upon by a number of forces. It has several parts involving AO1, AO2 and AO3, with verbal and mathematical skills. It starts with a recall part (a), on the conditions for equilibrium of an object under the influence of a number of forces. This part is designed to point you in the right direction for the analysis and evaluation of part (b).

Mark scheme

Question part			Description	AOs			Total	Skills	
				1	2	3		M	P
(a)			1. The [vector] sum of the forces on the object is zero or equivalent. [1] 2. The sum of the clockwise moments *about any point* equals the sum of the anticlockwise moments *about that point*. Non italic: [1], italic: [1]	3			3		
(b)	(i)	(I)	Vertical downward arrow from centre of rod, labelled 11.8 N (unit needed)[1] Vertical downward arrow on the spotlight, labelled 35.3 N (unit needed) [1]		2		2	2	
		(II)	Correct equation from applying Principle of moments, **or** by implication ecf on 11.8 N and 35.3 N [1] 2nd correct eqn, e.g. $T_A + T_B = 11.8 + 35.3$ or PoM about a 2nd point. [1] $T_A = 14.0$ N (accept 14 N) [1] $T_B = 33.1$ N (accept 33 N) [1] Ecf on second force to be calculated if wrong only because of mistake in first.		4		4	4	

	(ii)	Statement (in words): **Either**: least mass if hung from end of bar **Or**: Wire further from mass will be slack (or equiv) when rod about to tilt [1] (If hung from left): $mg \times 0.2$ [m] $= 11.8$ [N] $\times 1.3$ [m] [1] $m = 7.82$ kg [1]			3		1	
						3	1	
Total			3	6	3	12	8	

Rhodri's answers

(a) The sum of the upward forces on the object is equal to the sum of the downward forces. (not enough)

The sum of the clockwise moments equals the sum of the anticlockwise moments. ✓ ✗

MARKER NOTE

We need force components in any direction to sum to zero. First mark not gained.

Third mark not gained as it's important to state that the clockwise and anticlockwise moments are taken about the same point.

1 mark

(b) (i) (I)

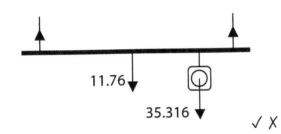

11.76

35.316

✓ ✗

MARKER NOTE

The second mark is lost for the omission of the unit. Rhodri has given too many significant figures, but this is more likely to be penalised in a question based on experimental results. Best avoided though!

1 mark

(II) Moments about B
$11.8 \times 1.3 + 35.3 \times 0.6 = T_A$ ✗
$\therefore T_A = 36.5$ N ✗
$T_A + T_B = 11.8 + 35.3$ ✓ecf
$\therefore T_A + T_B = 47.1$
$\therefore T_B = 47.1 - 36.5 = 10.6$ N   ✓ecf

MARKER NOTE

In the first equation Rhodri has written T_A instead of 2.6 [m] $\times T_A$. Writing a force instead of a moment is an error of principle (and a surprisingly common one), so the first and third marks are lost. However, $T_A + T_B = 11.8 + 35.3$ is the correct force equation and is used correctly to find T_B, with ecf on T_A. So second and fourth marks are gained.

2 marks

(ii) Tilting most likely if mass hung from end   ✓
$mg \times 0.2 = 11.8 \times 1.3 + 10.6 \times 2.6$ ✗
mass $= 22$ kg   ✗

MARKER NOTE

Only the first mark awarded for the realisation of where the mass needed to be hung. Rhodri did not realise that the tension in the far wire would be zero and (in desperation?) used the previous value.

1 mark

TOTAL   **5 marks / 12**

Ffion's answers

(a) The sum of the forces in any one direction on the object is zero. ✓bod

The sum of the clockwise moments about the pivot is equal to the sum of anticlockwise moments ✓ about the pivot. ✗

MARKER NOTE

Ffion should have written 'the sum of the <u>components</u> of the forces in any direction is zero'. This is a subtle point and the examiner awarded the first mark bod.

The third mark was not awarded because there does not have to be a pivot — moments can be taken about <u>any</u> point.

2 marks

(b) (i) (I)

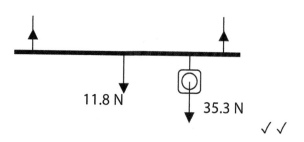

11.8 N   35.3 N   ✓✓

MARKER NOTE

A perfect answer.

2 marks

(II) $1.3T_A + 24.71 = 1.3T_B$

$T_A + T_B = 47.1 \, N$ ✓

$\therefore 1.3T_A + 24.71 = 1.3 \, (47.1 - T_A)$

$\therefore 2.6T_A = 46.52$

$\therefore T_A = 17.9 \, N$ and $T_B = 29.2 \, N$ ✓✗✓

MARKER NOTE

The first equation resulted from applying the PoM about the centre of the rod. Since the candidate did not tell us this, any slip in calculating the 24.7(1) N m moment of the spotlight would have made the method almost impossible to follow and most of the marks would have been lost — a high-risk tactic! Note also that it would have been neater to take moments about A or B because the 'perpendicular distance' of T_A or T_B would be zero so only one of the unknown tensions would appear in the equation. As it is, the candidate has solved a pair of simultaneous equations, making only an arithmetical mistake (46.52 instead of 36.52), and losing the third mark.

3 marks

(ii) $0.2 W = 1.3 \times 11.8$ ✗✓

$\therefore W = 76.7 \, N$

$mass = \dfrac{76.7}{9.81} = 7.8 \, kg$ ✓

MARKER NOTE

The first mark is lost because the candidate did not make any attempt to explain her reasoning, as required in the question. The analysis was spot on.

2 marks

TOTAL   **9 marks / 12**

Section 2: Kinematics

Topic summary

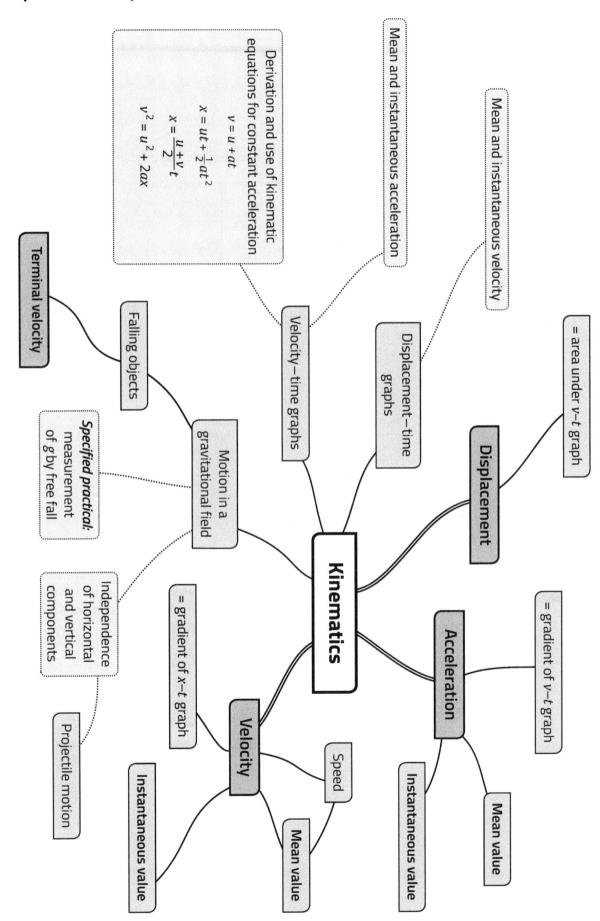

Derivation and use of kinematic equations for constant acceleration

$$v = u + at$$
$$x = ut + \frac{1}{2}at^2$$
$$x = \frac{u+v}{2}t$$
$$v^2 = u^2 + 2ax$$

Mean and instantaneous acceleration

Mean and instantaneous velocity

Terminal velocity

Falling objects

Velocity–time graphs

Displacement–time graphs

Displacement

= area under v–t graph

Specified practical: measurement of g by free fall

Motion in a gravitational field

Kinematics

Acceleration

= gradient of v–t graph

Independence of horizontal and vertical components

= gradient of x–t graph

Velocity

Speed

Instantaneous value

Mean value

Instantaneous value

Mean value

Projectile motion

Instantaneous value

Q1 (a) Define:

(i) Mean speed. [1]

....

....

(ii) Mean velocity. [1]

....

....

(b) Rhiannon runs from A to C around two sides of a square field (see diagram). She takes 27 s. Calculate:

(i) Her mean speed. [1]

....

....

(ii) Her mean velocity. [2]

....

....

Q2 A ball moving to the east at 19 m s^{-1} collides with a vertical wall and bounces back in the opposite direction at 11 m s^{-1}. The time of contact with the wall is 25 ms. Determine the ball's mean acceleration during the collision, and comment on its magnitude. [4]

....

....

....

....

....

....

....

Q3 (a) Derive the following equations for uniformly accelerated motion. If you choose to do so, you may use the sketch graph, adding your own labels.

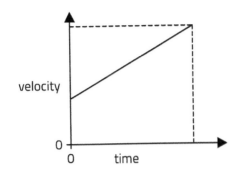

(i) $v = u + at$ [1]

....

....

....

(ii) $x = \dfrac{u + v}{2}t$ [2]

....

....

....

(iii) $x = ut + \frac{1}{2}at^2$ [2]

....

....

....

....

(b) Use the equations $v = u + at$ and $x = \dfrac{u + v}{2}t$

to derive the following equation for uniformly accelerated motion.

$$v^2 = u^2 + 2ax.$$ [3]

....

....

....

....

Q4 A stone is thrown with a velocity of 15.5 m s^{-1} upwards.

(a) Neglecting forces other than the Earth's gravitational pull, determine

(i) its maximum height above its point of launch, [2]

....

....

....

(ii) the time it takes to reach its maximum height. [2]

....

....

....

(b) Bryn says that the time taken for the ball to reach half its maximum height should be half the time to reach its maximum height. Without further calculation, evaluate this claim. [2]

....

....

....

(c) (i) Calculate the time it takes **from launch** for the stone to be at half its maximum height, on the way back down. [3]

....

....

....

....

(ii) Calculate its speed at this time. [2]

....

....

....

Q5 The Ilyushin Il-76 is an aeroplane that can be used as a water bomber. It carries a large quantity of water for dropping on to burning areas, e.g. forest fires. A bomber is flying at a height of 100 m at a speed of 120 m s^{-1}. It releases its payload (the water) before it is over the fire.

(a) Explain in terms of the motion of the water why it has to do this. [2]

....

....

(b) Calculate how far before the burning area the plane should release the water. [Ignore the effect of air resistance.] [3]

....

....

....

....

....

Q6 Huw throws a ball at a speed, u, of 20.0 m s^{-1}, at an angle, θ, of 37° above the horizontal.

(a) (i) Calculate its initial horizontal velocity component. [1]

....

....

(ii) Calculate its initial vertical velocity component. [2]

....

....

(iii) Huw is worried that his answers to (a) (i) and (ii) add up to more than the original speed.
Discuss whether or not he ought to be worried. [2]

....

....

....

(b) Determine:

(i) The maximum height gained by the ball. [2]

....

....

....

(ii) The ball's horizontal range (from its point of launch to its return to the same level). [3]

....

....

....

....

(c) Huw reads in an A level maths textbook that the range, R, of a projectile is given by:

$$R = \frac{u^2 \sin 2\theta}{g}$$

(i) Show that the equation is homogeneous (i.e. it works in terms of units). [2]

....

....

....

(ii) Compare the result of using this equation with your answer to (b)(ii). [1]

....

....

....

Q7 A velocity–time graph is given for a cyclist on a straight road.

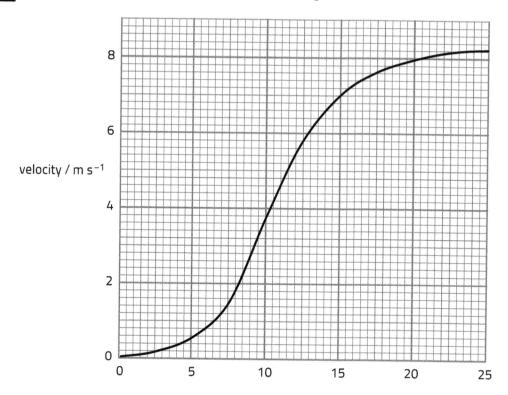

(a) Adding to the graph if you wish to do so, determine:

(i) The magnitude of the cyclist's mean acceleration between 10.0 s and 20.0 s. [2]

....

....

....

(ii) The magnitude of her acceleration at 15.0 s. [4]

....

....

....

....

....

(b) Iolo claims that (a) (ii) could be answered by determining the mean acceleration over the time interval 14.5 s to 15.5 s. Evaluate his claim. [2]

....

....

....

....

Q8   An idealised velocity–time graph for an electric car travelling between two sets of traffic lights is as follows:

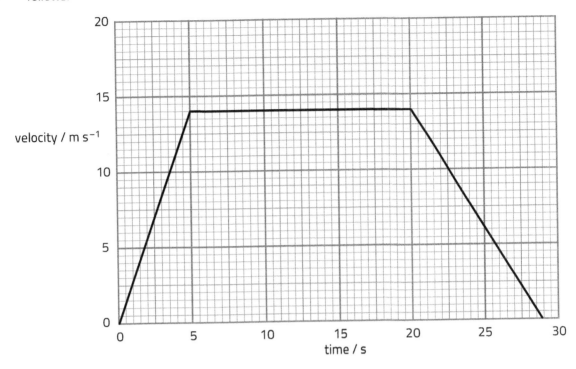

(a)   Determine the mean velocity of the car during the motion.   [3]

....

....

....

....

....

(b)   Using the same grid, draw a displacement–time graph for the motion. You will need to add a displacement scale to the grid.   [3]

Space for additional calculations, if required.

(c)   Sketch an acceleration–time graph for the motion in the space below. Label significant values.   [3]

Q9 Helen is sitting in the open carriage of a narrow-gauge railway which is travelling at a constant speed of 8.0 m s⁻¹. She throws a ball vertically upwards (from her point of view) at a speed of 10.0 m s⁻¹ (i.e. the actual vertical component of the ball's velocity is 10.0 m s⁻¹) and then catches it.

You should ignore the effects of air resistance for parts (a) – (c) of this question.

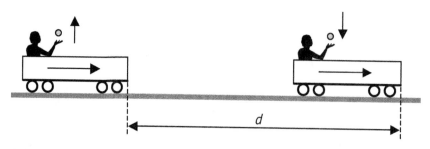

(a) Calculate the distance that the train moves during the time the ball is in the air. [3]

....

....

....

....

....

....

(b) Add to the diagram above to sketch the path taken by the ball during its motion through the air, as seen by an observer standing outside the train. [1]

(c) Explain in terms of the vertical and horizontal motions of the ball why Helen is able to catch the ball in spite of having moved the distance you calculated in part (a). [2]

....

....

....

....

(d) Discuss how air resistance would have affected the motion of the ball. Think about it from the points of view of Helen and the observer. [3]

....

....

....

....

....

....

Question and mock answer analysis

Q&A 1 A velocity–time graph (with scales removed from the axes) is given for a body moving along a vertical line. Describe what the graph tells us about the body's motion during each stage, AB, BC and CD, in terms of its velocity, its acceleration and its displacement from its starting point (at A).

[6 QER]

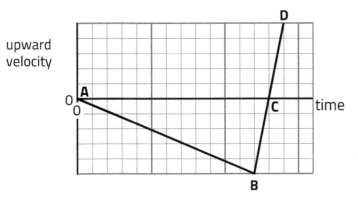

What is being asked

On the face of it, this is a very straightforward question about the interpretation of a velocity–time graph. The thing to ensure, in this QER question, is that all the points are covered, i.e. the sections AB, BC and CD as well as the three aspects of the motion, and all in a well-linked narrative. For a mark in the top band (5–6 marks), you need give a comprehensive description of velocity, acceleration and displacement in all three stages. The middle (3–4 marks) and bottom (1–2 marks) bands have less good coverage of aspects of motion or of the stages, or both.

Mark scheme

Question part			Description	AOs			Total	Skills	
				1	2	3		M	P

			Indicative content
			 AB – velocity: increases from zero; downwards
			     acceleration: constant; downwards
			     displacement: increases downwards; at an
			       increasing rate
			 BC – velocity: downwards; falls to zero
			     acceleration: constant; upwards; large compared to
			       over AB
			     displacement: more downward displacement; at a
			       decreasing rate; only a little more
			 CD – velocity: upwards; increases from zero; ends at the
			       reverse velocity as B
			     acceleration: same as over BC
			     displacement: decrease in downward displacement;
			       but only small / equal to gain over BC.
				2	4		6		

Total				2	4		6		

Rhodri's answers

In AB the body moves downwards, getting faster, therefore it accelerates. Its displacement (down) steadily increases.

In BC the body loses all its velocity. It has a fast acceleration up. Its displacement is still increasing.

In CD the body gets faster again, but going up, with the same acceleration as before. The downward displacement decreases because the body is going up.

MARKER NOTE

There are no glaring mistakes in this answer, but there is a general lack of detail, which makes this a middle-band answer. In AB we ought to have been told that the acceleration is downwards. 'Steadily' is not a good word to describe how the displacement is increasing.

In BC one minor criticism is that accelerations aren't *fast* or *slow*, but *large* or *small*, *high* or *low*. It was not mentioned that the extra displacement is small compared with that over AB.

In CD 'the same acceleration as before' is ambiguous. We also needed something to be said about how much or how little the downward displacement decreases.

4 marks

[**Comment:** Marking QER questions involves judgement. A different examiner might have awarded 3 marks.]

Ffion's answers

Over AB the body's velocity decreases, but actually the body is getting faster. This means that it has an acceleration that is negative and constant (constant gradient). Its displacement is negative and increasing in amount.

Over BC the body is slowing down, though the slope of the graph is large and positive so the acceleration is positive, constant and larger than the acceleration over AB. The negative displacement is still increasing in amount but not by very much as the area under the graph is small here.

Over CD the body's velocity is now increasing. The acceleration is the same as in BC. The body's negative displacement decreases a little.

MARKER NOTE

This candidate's answer would have been excellent if she had used *upwards* instead of *positive* and *downwards* instead of *negative*. The rather confused first sentence suggests that some conceptual problem about vector quantities. It would have been clearer if she had said that the magnitude of the velocity increases in AB.

One small omission is that no comment has been made about rates at which displacement changes in any stage.

The question required descriptive statements rather than explanations, but the candidate has justified her answer (for BC) by mentioning graph gradient and area under graph. These two short remarks do tend to enhance her answer, but too much explanatory material would have clogged up the description asked for.

The good coverage of a, v and x in all three regions makes this a top-band answer.

5 marks

Q&A 2

(a) In an experiment to determine a value for g, a student drops six marbles, one at a time, from an eighth-floor window of a tall building. Another student uses a stopwatch to time how long they take to reach the ground (in a roped-off area). The height dropped is measured as 24.30 m ± 0.01 m. The times recorded are:

2.31 s 2.47 s 2.26 s 2.42 s 2.35 s 2.51 s

Calculate a value for g from these results, along with its absolute uncertainty, giving your working.

[6]

(b) Comment on the value for g and its uncertainty found in part (a), and briefly discuss likely causes of error and uncertainty. [4]

Unit 1 Practice questions

What is being asked

This question directly tackles experimental skills. It is set around one of the set practicals: the measurement of g by free fall. Part (a) involves a standard use of experimental data to determine a result with its associated uncertainty. The uncertainty in one variable (the drop height) is given; the uncertainty in time must be calculated from the raw results. Part (b) requires an evaluation of the final result.

Mark scheme

| Question part | Description | AOs | | | Total | Skills | |
		1	2	3		M	P
(a)	mean time = 2.39 s [1] $g = \dfrac{2h}{t^2}$ (or by implication) [1] $g = 8.51$ m s^{-2} or 8.5 m s^{-2} [1] $\Delta t = \dfrac{2.51 - 2.26}{2}$ (or by impl.) [1] [= ± 0.13 s] $p_g = 2 \times \dfrac{0.13\,(\text{ecf})}{2.39} \times 100$ or by impl. [1] [= 11%] $\Delta g = \pm\, 0.9$ m s^{-2} **or** 0.94 m s^{-2} if g to 2 d.p. [1] **Alternative for last 3 marks** Either max or min g calculated [9.52 m s^{-2}; 7.71 m s^{-2}] ✓ g uncertainty = (max – min) / 2 , or equiv. ✓ $\Delta g = \pm\, 0.9$ m s^{-2} **or** 0.91 m s^{-2} if g to 2 d.p. ✓		6			4	6
(b)	The value for g is low or the [%] uncertainty is large [1] The standard value of g isn't allowed by the uncertainty [1] Large [random] uncertainty expected when timing such short intervals by eye with stop-watch. [1] Low value of g probably due to a [systematic] error, e.g. air resistance [over long fall] **or** in timings. [1]			4			4	4
Total			6	4	10	8	10

Rhodri's answers

(a) $2.31 + 2.47 + 2.26 + 2.42 + 2.35 + 2.51 = 14.32$

$14.32 \div 6 = 2.39$ ✓

$24.3 = \frac{1}{2} \times g \times 2.39^2$

$g = \frac{2 \times 24.3}{2.39^2}$ ✓ $= 8.5 \text{ m s}^{-2}$ ✓

$2.51 - 2.31 = 0.2$ ✗

$0.2 \div 2 = 0.1$, $\frac{0.1 \times 100}{2.39} = 4.18\%$

$4.18\% \times 2 = 8.36\%$ ✓ ecf

$\frac{9.51 - 8.51}{2} = \pm 0.71 \text{ m s}^{-2}$ in g ✗

MARKER NOTE

Rhodri has followed a correct procedure, showing all his steps. It would have been safer, though, to give a few more *words* stating what his figures represent. He has made a slip in calculating his uncertainty in the times, choosing 2.31 s instead of 2.26 s as his lowest; fourth mark lost. Since he has given g to only 1 dp, it is wrong to give its uncertainty to 2 dp; sixth mark lost.

4 marks

(b) The value of g is too low, since the right value is 9.8 m s^{-2}. ✓ This may be because of air resistance (drag). ✓ A stopwatch isn't good enough for such short time intervals. (not enough)

MARKER NOTE

Rhodri's first sentence gains the first (easy) mark, and his second sentence, the fourth mark. He's missed the second mark, not commenting that his range of allowed values doesn't include 9.81 m s^{-2}. He realised that the stop-watch technique isn't really suitable, but hasn't linked it with uncertainties, so misses the third mark.

2 marks

Total   **6 marks / 10**

Ffion's answers

(a) Mean of times =

$\frac{2.31 + 2.47 + 2.26 + 2.42 + 2.36 + 2.51}{6} = 2.39 \text{ s}$ ✓

$x = ut + \frac{1}{2}at^2$

$u = 0 \rightarrow a = \frac{2x}{t^2} = \frac{2 \times 24.3}{2.39^2}$ ✓ $= 8.5 \text{ m s}^{-2}$ ✓

Highest possible acceleration is for 2.26 s

so max $a = \frac{2x}{t^2} = \frac{2 \times 24.3}{2.26^2} = 9.51 \text{ m s}^{-2}$ ✓

Uncertainty $= \frac{9.51 - 8.51}{2}$ ✗ $= 0.5 \text{ m s}^{-2}$ ✗

(no ecf) ✗

MARKER NOTE

This candidate's answer is very clearly and competently set out. She has chosen the easy way to calculate the absolute uncertainty in g, but has made a serious error in the last step: the uncertainty is 9.51 − 8.51, not $\frac{9.51 - 8.51}{2}$.

Presumably she was getting confused with methods that do require division by 2. The last mark cannot be awarded ecf because the mistake is one of principle, and not just a slip. The last two marks are lost.

4 marks

(b) Even the highest possible acceleration is less than the accepted value of 9.81 m s^{-2}. ✓✓. This might be because the timekeeper stops the stop-clock too late every time, ✓ or it might be because air resistance really does reduce the acceleration. In any case, a 0.1 s error in timing such a small interval makes for a large percentage uncertainty. ✓ A stopwatch isn't good enough for such short time intervals.

MARKER NOTE

Although she has used different words from those of the mark scheme, Ffion's first sentence gains both of the first two marks. Her second sentence contains more than enough to give her the fourth mark. Her next sentence gains her the third mark. A competent answer.

4 marks

Total   **8 marks / 10**

Section 3: Dynamics

Topic summary

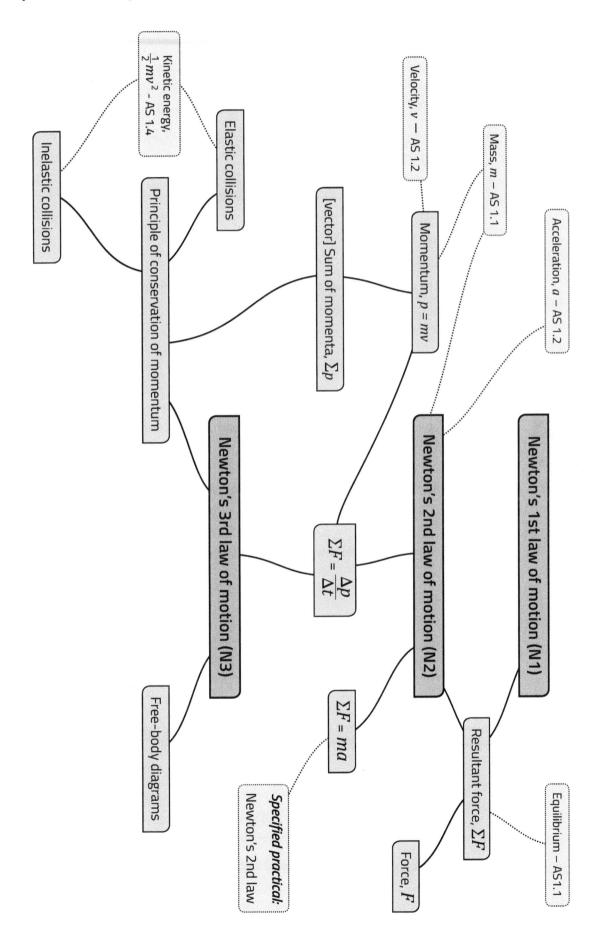

Q1 Newton's third law of motion is about pairs of forces. In part (b) of this question they will be called 'Newton's third law partner forces'.

(a) State Newton's third law of motion. [1]

....

....

....

(b) The diagram shows a book at rest on a table.

(i) **Show on the diagram** the two main forces on the book, using labelled arrows. [2]

(ii) For each of the two forces you have labelled, state the body on which the Newton's third law partner force acts. [2]

....

....

Q2 A teacher asks two students to write Newton's second law of motion as an equation. Bethan, a GCSE student, writes:

$$F = ma.$$

Angharad, an A level student, writes:

$$F = \frac{\Delta p}{t}.$$

(a) Defining all the symbols, explain how these two equations are consistent with each other. [3]

....

....

....

....

(b) The force that a gas exerts on the walls of its container arises because the gas molecules hit the walls and bounce off. Explain which of the two equations is more useful in this context. [2]

....

....

....

....

Q3 A box of mass 28 kg is dragged along level ground in an easterly direction by means of a rope.

The box has a constant acceleration. As it slides a distance of 2.7 m its velocity increases from 1.5 m s^{-1} to 2.1 m s^{-1}.

rope

(a) Calculate the resultant force on the block.   [3]

....

....

....

....

....

(b) During the acceleration the rope exerts a force of 18.2 N on the box. Determine the magnitude and direction of the frictional force that **the box exerts on the ground**, giving your reasoning.   [3]

....

....

....

....

....

Q4 Two trolleys, A and B, each of mass 12 kg, are linked by a chain of mass 2.0 kg. They are pulled by a rope and their acceleration is 0.75 m s^{-2}. The frictional force on each trolley is 5.0 N.

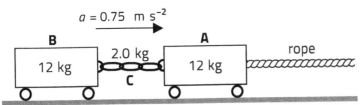

Explaining your reasoning, calculate:

(a) The force exerted by the rope on trolley **A**.   [2]

....

....

....

(b) The force exerted by trolley **B** on the chain, **C**.   [3]

....

....

....

....

Q5 A body of mass 4.0 kg is acted upon by forces in the horizontal plane, as shown.

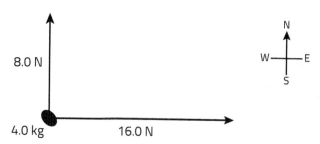

Determine the magnitude and direction (as a bearing) of the body's horizontal acceleration.
You may add to the diagram or draw a separate vector diagram in the space to the right of it. [4]

....

....

....

....

....

Q6 The diagram shows a steel ball attached by stretched springs to fixed anchorages. The ball is held in position by an electromagnet (not shown). The tension in each spring is 6.0 N.

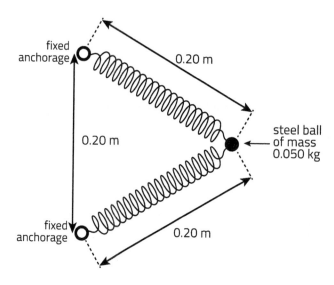

(a) Calculate the ball's acceleration just after the electromagnet is turned off. [4]

....

....

....

....

....

....

(b) Give two reasons why the ball's acceleration will decrease as the ball moves. [Resistive forces are negligible.]   [3]

....

....

....

....

Q7   An empty supermarket trolley is pushed and released so that it nests with two identical trolleys that are already nested and moving more slowly in the opposite direction (see diagram). After the collision, the three trolleys move off together with the same velocity.

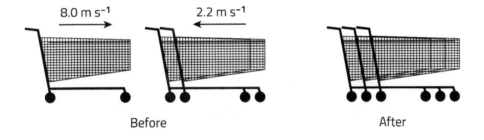

8.0 m s⁻¹   2.2 m s⁻¹

Before   After

(a) Determine the velocity of the trolleys after the collision.   [3]

....

....

....

....

(b) Express the total final kinetic energy of the three trolleys as a fraction of their total initial kinetic energy.   [3]

....

....

....

....

(c) A student claims that the 'lost' kinetic energy has been given to the atoms of the trolleys. Discuss this claim.   [2]

....

....

....

Q8 A dumbbell consists of two metal weights, each of mass 2.5 kg, separated by a light rod. The dumbbell is rotating about its centre, so that the speed of the weights is 5.0 m s^{-1}.

Calculate the total kinetic energy and the total momentum of the dumbbell. Explain your reasoning. [4]

....

....

....

....

....

....

Q9 A ball of mass 0.220 kg is dropped on to a floor from a height of 2.00 m. It bounces back up, reaching a maximum height of 1.20 m.

(a) Show that the ball's change in momentum during the bounce is roughly 2.5 N s. [5]

....

....

....

....

....

....

....

(b) The ball is in contact with the ground for 150 ms. Calculate the mean resultant force on the ball during the bounce. [2]

....

....

....

(c) Explain why the mean force exerted by the ground on the ball during the bounce must be roughly 2 N greater than the mean resultant force on the ball calculated in (b). [2]

....

....

....

Q10 Two gliders of unequal mass approach each other on a level air-track, as shown.

After the collision the 0.25 kg glider is moving with a velocity of 0.107 m s^{-1} to the right.

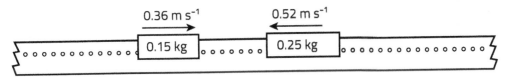

(a) Determine the velocity of the 0.15 kg glider after the collision. [3]

....

....

....

....

....

....

(b) Show that the collision is inelastic. [3]

....

....

....

....

....

....

Q11 A molecule of mass 6.6×10^{-27} kg travelling at 2 500 m s^{-1} bounces round the inside of an otherwise empty box as shown. The collisions with the walls are elastic and direction is always at 30° to side **XY**.

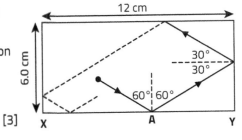

(a) Calculate the momentum change of the molecule at **A**. [3]

....

....

....

....

(b) Calculate the mean force the molecule exerts on the side XY. [Hint: Consider the time taken before the molecule hits the side **XY** again.] [3]

....

....

....

....

Question and mock answer analysis

Q&A 1 A cyclist, Carys, freewheels (rides without pedalling) down a hill. Carys and her bike (C-and-B) have a total mass of 78 kg, and are shown on the diagram as a single rectangle.

(a) **Add labelled arrows to the diagram** to show the forces acting on C-and-B. [3]

(b) Calculate the component of the gravitational force that accelerates C-and-B. [2]

(c) Carys's speed increases from 4.5 m s⁻¹ to 11.3 m s⁻¹ over a time of 12.0 s. Calculate the mean **resistive** force on C-and-B, giving your reasoning. [3]

(d) Carys believes that the acceleration must be greater towards the end of the 12.0 s interval than at the beginning. Evaluate this belief. [3]

What is being asked

The motion, or lack of it, of objects on inclined planes is a common setting for AS and A level questions. Knowledge from Sections 1.1 and 1.2 as well as 1.3 is needed here. In part (a), you are expected to remember that gravity (or weight) acts vertically downwards on an object, a flat surface always exerts a force at right angles on an object in contact and that friction will act in a manner which opposes motion. This is classed as AO1 even though the knowledge must be applied to this situation, as you will certainly have encountered the set-up before. Taking components in (b) is a skill from Section 1.1, and calculating acceleration, which is required in (c), is from 1.2, this time applied to forces and motion, a 1.3 concept. (b) and (c) are AO2 questions. Examiners very often pose a supposed student's statement and ask for comments; this is a standard way of setting an AO3 question. In this type of question, there is no mark for just saying that Carys is wrong (or right) but this is required as part of the answer.

Mark scheme

Question part		Description	AOs			Total	Skills	
			1	2	3		M	P
(a)		Downward arrow labelled *weight* (or equiv) [1] Arrow normally 'up' from surface labelled *normal contact force* (accept *normal reaction*) [1] Arrow up slope labelled *resistive force* or equiv [1] No penalties for oddly *positioned* arrows. [−1 mark lost per additional incorrect force]	3			3		
(b)		Multiplication by sin 6.0° [1] [78 × 9.81 × sin 6.0° =] 80 N [1]		2		2	2	
(c)		Mean acc = $\dfrac{11.3 - 4.5}{12.0}$ [m s^{-2}][1][= 0.567 m s^{-2}] Mean resultant force = 78 × 0.567 N [1] [= 44.2 N] Mean resistive force = [80 N − 44 N] = 36 N [1] Full ecf on 80 N.		3		3	1 1	
(d)		Resistive force [**or** air resistance] increases as speed increases [1] So resultant force decreases [1] So acceleration decreases **and** Carys wrong. [1]			3	3		
Total			3	5	3	11	4	

Rhodri's answers

(a)

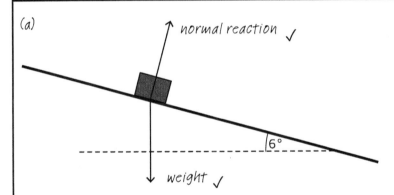

normal reaction ✓

6°

weight ✓

MARKER NOTE
Rhodri gains marks for correctly marking on the weight and 'normal reaction'. He omits the resistive force and so doesn't gain the third mark.

Rhodri has made no attempt to put the tails of the arrows at the points where the forces act, but it is not easy to do so in this case, and there was no penalty for not doing so.
2 marks

(b) Weight = 78 × 9.81 = 765.18 N

Comp down slope = 765.18 × cos 6° ✗

= 761 N ✗ no ecf

MARKER NOTE
Rhodri has multiplied the weight by cos 6° rather than by sin 6° (or cos 84°). This is a serious error and there is no separate mark for calculating *mg*.
0 marks

(c) Change in velocity = 11.3 − 4.5

= 6.8 m s^{-1}

so accel $\dfrac{6.8}{12}$ = 0.56666 m s^{-2} ✓

so force = 78 × 0.56666 = 44.2 N ✓ bod

MARKER NOTE
Rhodri has not calculated the resistive force. He correctly calculated the resultant force, but has not stated that it *is* the resultant force. Examiner gave benefit of doubt here and so he obtains the first two marks.
2 marks

(d) Carys is obviously right. She gets faster and faster as she goes down the slope. ✗

MARKER NOTE
Rhodri appears to be confusing acceleration with speed or velocity. No credit given.
0 marks

Total   **4 marks / 11**

Ffion's answers

(a)

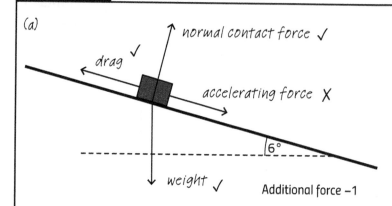

normal contact force ✓

drag ✓

accelerating force ✗

weight ✓

Additional force −1

MARKER NOTE
Ffion gains marks for correctly marking on the weight and the normal contact force (a preferable term to 'normal reaction'). She also marks the weight but also includes an 'accelerating force' which is not a separate force but is either a component of the weight or the resultant of this component and the drag. This arrow shouldn't be there and she is penalised 1 mark.

2 marks

(b) Weight = 78 × 9.81

accelerating force = 78 × 9.81 × sin 6° ✓

= 80 N ✓

MARKER NOTE
Clear and correct. Both marks obtained.

2 marks

(c) Acc'n due to 80 N = $\frac{80}{78}$ = 1.03 m s^{-2} ?

actual acc'n = $\frac{6.8}{12}$ = 0.57 m s^{-2} ✓

Therefore dec'n due to resistive force = 1.03 − 0.57 = 0.46 m s^{-2} ?

Thus resistive force = 78 × 0.46 = 36 N ✓

MARKER NOTE
It is no co-incidence that Ffion's procedure gives the right answer. Yet only one of her three accelerations (the 'actual acceleration') has any physical meaning. The reasoning should have been conducted in terms of forces. 1 mark withheld.

2 marks

(d) She is going fastest at the end, so air resistance is greatest then ✓ so Carys is wrong. ✗

MARKER NOTE
The 3 marks allotted suggest that a full answer is needed. Ffion's, though it starts well, lacks the middle step of considering the resultant force. There is no mark for just saying that Carys is wrong.

1 mark

| Total | 7 marks / 11 |

Section 4: Energy concepts

Topic summary

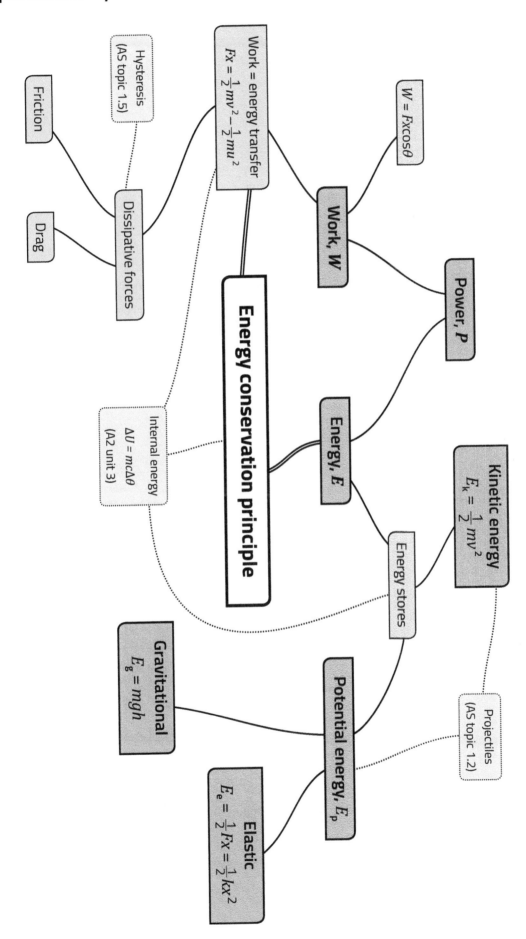

Q1 The kilowatt-hour (kW h) is a unit of energy.

(a) Use the definition of power to explain why the kW h is a unit of energy. [1]

....

....

(b) The battery of an electric car is rated as 96 kW h. Express this in joule (J) in standard form. [2]

....

....

....

Q2 (a) State the principle of conservation of energy. [1]

....

....

(b) A stone of mass 0.150 kg is thrown diagonally upwards with a speed of 50.0 m s^{-1}.

(i) Assuming that air resistance is negligible, use the principle of conservation of energy to calculate the stone's speed at its maximum height of 31.9 m. [3]

....

....

....

....

....

(ii) Without calculation, explain whether the answer to (i) would be different if the mass of the stone were different. [2]

....

....

....

Q3 A cyclist freewheels down a hill which has a gradient of 5.0°. After travelling 200 m from rest, her speed is 12.0 m s^{-1}. The combined mass of the cyclist and bicycle is 85 kg.

(a) Calculate the loss in gravitational potential energy. [2]

....

....

....

(b) Using an energy argument, calculate the mean resistive force acting. [2]

....

....

....

Q4   The Highway Code gives the typical braking distance for a car travelling at a speed of 60 mph (26.7 m s^{-1}) on a dry road with good tyres as 55 m.

(a)   Calculate the braking force assumed by the Highway Code for a car of mass 1200 kg.   [3]

....

....

....

....

....

(b)   The braking distance for cars travelling at 30 mph, 50 mph and 70 mph are given as 14 m, 38 m and 75 m respectively. Evaluate whether the Highway Code assumes a constant braking force.   [3]

....

....

....

....

....

(c)   The typical braking distance figures given in the Highway Code are the same for all cars. John states that this means that all cars are assumed to have the same braking force. Discuss whether John's statement is correct.   [2]

....

....

....

Q5   The WJEC Terms and Definitions booklet defines power as 'the work done per second or the energy transferred per second'.

(a)   Use a definition of work or energy to explain why the two alternative parts of the definition are equivalent.   [2]

....

....

....

(b)   John states that in some situations either alternative is useful but sometimes only one of them is appropriate.

The Sun's power output is approximately 3.8 × 10^{24} W; a horse dragging a log has a useful power output of about 600 W. Use these examples to illustrate John's statement.   [3]

....

....

....

....

....

Q6 A vertical spring with a spring constant 32.0 N m^{-1} is clamped at its top end. A load of mass 0.600 kg is attached to the bottom of the spring, at a height of 0.400 m above the laboratory bench and supported so that the spring is not stretched. The load is then released.

(a) Show that the initial gravitational potential energy (from bench height) is approximately 2.4 J. [2]

....

....

....

(b) Calculate the loss in gravitational potential energy, the elastic potential energy and the kinetic energy when the load has fallen 0.184 m. [3]

....

....

....

....

(c) Use your answer to (b) to calculate the speed of the load when it has fallen 0.184 m. [2]

....

....

....

(d) Show that the lowest point that the load reaches is 0.032 m above the bench. [3]

....

....

....

....

....

(e) A student correctly works out that the speed, v, of the load when it has fallen by a distance h can be worked out using the equation:

$$mgh = \tfrac{1}{2}kh^2 + \tfrac{1}{2}mv^2$$

(i) Explain this equation in terms of energy transfer. [2]

....

....

....

(ii) Use the equation above to calculate the values of h at which $v = 1.00$ m s^{-1}. [3]

....

....

....

....

Q7 (a) The work, W, done by a force of magnitude, F, which moves its point of application by a distance d in the same direction as F is given by $W = Fd$.

Use definitions of velocity and power to show that the power, P, transferred by a force F moving at a velocity v is given by $P = Fv$. [2]

....

....

....

....

(b) The aerodynamic drag, D, on a new-model electric SUV is given by $D = kv^2$, where k is a constant, with a value of 0.4 in SI units.

(i) The unit of k can be written N (m s^{-1})$^{-2}$. Express this in base SI units in the simplest terms. [2]

....

....

....

....

(ii) Calculate the power transferred by the SUV's engine when the car is travelling at a steady speed of 30 m s^{-1}. [3]

....

....

....

....

....

....

(iii) An advert for the SUV says that, with a 100 kW h battery, the range of the car is 900 km. Evaluate this claim. [You should make reasonable assumptions about the efficiency of the engine drive system and the driving speed.] [4]

....

....

....

....

....

....

....

....

Q8 A traveller in the Arctic drags a laden sledge of total mass 210 kg a distance of 7.0 km across a frozen surface, in a straight line at a constant speed of 1.4 m s^{-1}. To do so, he applies a constant force of 83 N to the sledge via a tow-rope, which is inclined at an angle of less than 5° to the horizontal.

(a) Calculate the work done on the sledge by the traveller. [2]

....

....

....

(b) Considering the number of significant figures in the data, explain why, to answer part (a), it was not necessary to know the actual angle of the tow-rope to the horizontal. [2]

....

....

....

(c) State the value of the frictional force between the sledge and the ice. [1]

....

(d) For most of the 7.0 km journey, the sledge gains no kinetic energy. Explain why not and state the nature of the energy transferred to the sledge and ice. [2]

....

....

....

(e) At the end of the journey, the friction between the sledge and the ice causes the sledge to come to a stop.

(i) State the energy change that occurs during this process. [1]

....

(ii) Calculate the distance travelled by the sledge in stopping. [2]

....

....

....

(f) At the start of the journey, the traveller applies a force of 105 N to the sledge to accelerate it from rest to its steady speed. Aled claimed that most of the work done in the process went into the kinetic energy of the sledge. Evaluate this claim. [2]

....

....

....

Q9 Bethan, a hill farmer in mid-Wales, installs a small electrical wind turbine, with 6.0 m blades, to provide power for her farm. On one particular day, the wind speed is 15 m s^{-1}.

(a) She calculates that the mass, m, of air interacting with the blades of the wind turbine per second is given by:

$$m = \pi r^2 v \rho$$

where r is the radius of the blades, v is the wind speed and ρ the density of the air.

(i) Show that this equation is homogeneous, i.e. that the units are the same on the two sides. [2]

....

....

....

(ii) By considering the kinetic energy of the air, show that the power input, P_{IN}, to the wind turbine can be calculated using the formula:

$$P_{IN} = \tfrac{1}{2} \pi r^2 \rho v^3$$

[2]

....

....

....

(iii) The efficiency of the wind turbine itself is 56%. It is coupled to an electrical generator of efficiency 95% via a gear-box of efficiency 85%. Calculate the electrical power output, in kW, of the generator, to an appropriate number of significant figures.
[ρ_{air} = 1.3 kg m^{-3}] [4]

....

....

....

....

....

....

(b) Bethan knows that the output from the wind turbine is variable so plans to install rechargeable batteries to store energy for use in non-windy conditions, removing the need to buy electricity from the grid. Her mean power requirement is 75% of the value you calculated in (a)(iii). She knows that the mean wind speed on her farm is 7.5 m s^{-1}. She believes that the mean power output of the generator is therefore half the value that you calculated in (a)(iii), which would not be enough to provide all the power needs of the farm. Evaluate whether she is correct. [3]

....

....

....

....

....

Unit 1 Practice questions

Question and mock answer analysis

Q&A 1

(a) The diagram shows a steel sphere resting against a compressed spring. The knob is released, allowing the spring to expand quickly, causing the ball to travel along the track from A, upwards to D and horizontally towards E.

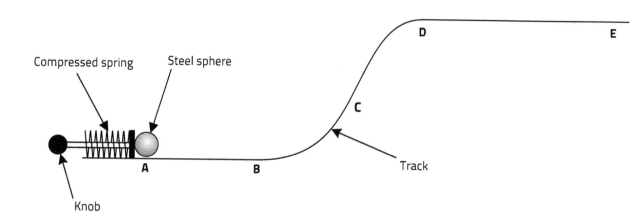

Explain the motion of the sphere in terms of forces and the energy transfers that occur from the moment the spring starts to expand. [6QER]

(b) A zip-wire ride in an adventure park has a length of 200 m and a fall of 25 m.

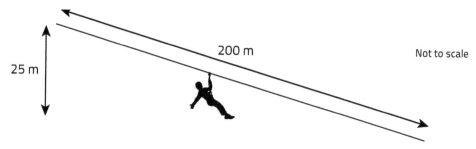

A rider of mass 70 kg achieves a speed of 15 m s^{-1} just before reaching the bottom of the ride.

Calculate the mean resistive force on the rider during the ride. [4]

What is being asked

This question is about energy transfers.

Part (a) is a *Quality of Extended Response* (QER) question, which is marked in terms of both content and clarity of the answer. It has both AO1 and AO2 aspects. The examiner is looking for an explanation of the forces that do work on the ball, causing named energy transfers. See page 8 for a general description of QER questions. In this case, examiners are looking for both knowledge of the energy stores (gravitational potential, elastic potential, kinetic, etc.) and the mechanism by which the transfer occurs, i.e. work by an identified force. Low-level answers (1–2 marks or 3–4 marks) would miss out some of the energy transfers, not appreciate that work is the means by which the energy transfers occur or incompletely identify the forces involved.

Part (b) is a calculation question following on from the theme of work and energy transfers in part (a), which is also AO1 and AO2, involving basic mathematical reasoning.

Mark scheme

Question part			Description	AOs			Total	Skills	
				1	2	3		M	P
(a)			**Indicative content** • Energy transfers • Identification of energy forms • Elastic → kinetic • Kinetic → gravitational potential • Energy loss as internal energy / heat • Transfers linked to positions **Forces and energy transfers** • Work as the agency of energy transfer • Force from spring: elastic → kinetic • Gravitational force: kinetic → gravitational • Dissipative forces: kinetic → internal / heat • Friction or air resistance as dissipative force • Molecular explanation of friction or air resistance.	2	4		6		
(b)			mgh used to calculate loss in E_p or by impl. [e.g. 17167.5 (J) seen] [1] $\frac{1}{2}mv^2$ used to calculate gain in E_k or by impl. [e.g. 7875 (J) seen] [1] Total loss in $E_p + E_k$ calculated, ecf, or by impl. [e.g. 9292.5 (J) seen] [1] Resistive force $= \left[\frac{9292.5}{200}\right] = 46(.5)$ N ecf [1]	2	2		4	4	
Total				4	6	0	10	4	

Rhodri's answers

(a) At the start, the energy is all potential energy in the spring. When the spring is released, it pushes the ball, giving it kinetic energy. As it moves, the ball has kinetic energy between A and B, then it gains potential energy and loses kinetic energy to D. After D it slows down, losing kinetic energy because of friction, until it stops. Then it just has potential energy.

(b) $PE = 70 \times 9.81 \times 25 = 17167.5$ J ✓

$KE = \frac{1}{2} \times 70 \times 15^2 = 7875$ J ✓

$v^2 = u^2 + 2ax$, $15^2 = 0^2 + 2a \times 200$,

so $a = \frac{15^2}{2 \times 200} = 0.5625$ m s^{-2} ✗

$F = ma = 70 \times 0.5625 = 39.375$ N ✗

no ecf

MARKER NOTE

Rhodri identifies most of the energy forms, omitting internal energy / heat, but doesn't distinguish between elastic and gravitational potential energy. He mentions two forces (spring push and friction) but does not mention the role of work in transferring energy. This is a low-level (bottom band) answer. His answer would be improved by identifying work against the force of gravity in the transfer from kinetic to gravitational potential energy and similarly work against friction in the transfer to internal / heat energy.

2 marks

MARKER NOTE

Rhodri starts well. He uses the hint of part (a) well and correctly calculates the change in gravitational potential and kinetic energies and so gains the first two marks. He doesn't explicitly state mgh or $\frac{1}{2}mv^2$ but he has clearly used them, as required in the mark scheme. There is no penalty for significant figures.

The remaining two marks are not given as the work is not relevant. Rhodri has in fact calculated the mean resultant force on the rider. There is an alternative method of answering the question, which involves Rhodri's calculation:

1 Calculate the component of the gravitational force down the wire (assumed straight) using $mg\sin\theta$ (85.8 N).

2 Subtract the resultant force Rhodri has calculated to give the 46.4 N, which is the correct answer.

However Rhodri cannot obtain marks for two partial answers.

2 marks

Total	4 marks / 10

Ffion's answers

(a) The forms of energy involved are: elastic potential energy (EPE), gravitational potential energy (GPE), kinetic energy (KE) and heat (H). To start with the spring has EPE because it is compressed. When it expands, it exerts a force on the ball, which does work and the ball gains KE (between A and B). Between B and D, the ball rises so it gains GPE and loses KE. The loss in KE is equal to the gain in GPE. So at D the ball has less KE than at B. As it rolls from D towards E it slows down because it hits air molecules (i.e. air resistance), so it loses more KE and the air gains heat energy (H).

MARKER NOTE

Ffion correctly identifies all the relevant stores of energy. In an A2 question, internal energy would be expected instead of heat but this is accepted here. She also correctly identifies work as the agency of energy transfer and, in two of the transfers, names the force involved. The role of air molecules as applying the force of air resistance is a good point to make.

The answer could be improved in two ways:

- By identifying the role of gravity in the transfer of kinetic energy to gravitational potential energy, i.e. work done against gravity.
- By noting that air resistance (and rolling friction) is significant all through the journey, not just after D.

This is a high-level answer.

5 marks

(b) Energy loss $= mgh - \frac{1}{2}mv^2$

$= 70 \times 9.81 \times 25 \checkmark - \frac{1}{2} \times 70 \times 15^2$

$= 1420 \text{ J (3 sf)}$ ✗ ✓ ecf

So work against friction $= 1420 = F \times 200$

So $F = \frac{1420}{200} = 7.1 \text{ N}$ ✓ ecf

MARKER NOTE

This is almost a perfect answer, with just one slip, which has cost Ffion a mark.

Ffion sets out the method of calculating the energy loss concisely, involving both the gravitational potential energy (for which she gets the first mark) and the kinetic energy. Unfortunately she has made the slip of not using the $\frac{1}{2}$ in calculating the KE, so she has not used $\frac{1}{2}mv^2$ and hence, she misses the second mark.

However, she obtains the third mark for combining the change in PE and KE on the ecf principle, as the error was just an arithmetic slip. The last mark is also given on the ecf principle although Ffion's answer of 7.1 N is wrong..

3 marks

Total **8 marks / 10**

Section 5: Solids under stress

Topic summary

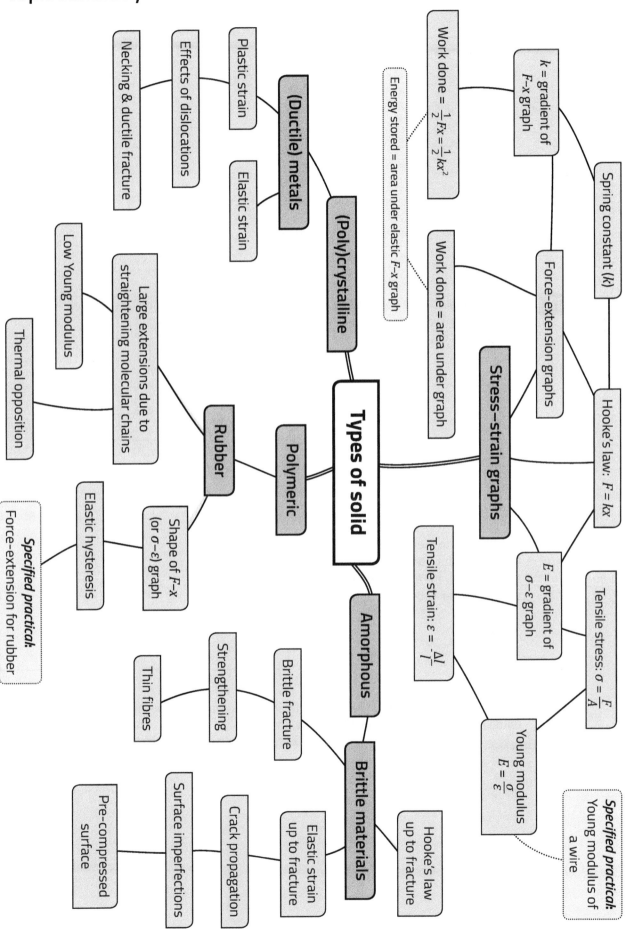

- Necking & ductile fracture
- Effects of dislocations
- Plastic strain
- **(Ductile) metals**
- Elastic strain
- **(Poly)crystalline**

- Work done $= \frac{1}{2}Fx = \frac{1}{2}kx^2$
- k = gradient of F–x graph
- Spring constant (k)
- *Energy stored = area under elastic F–x graph*
- Work done = area under graph
- **Force-extension graphs**
- Hooke's law: $F = kx$

- Low Young modulus
- Large extensions due to straightening molecular chains
- Thermal opposition
- **Rubber**
- **Polymeric**
- **Types of solid**
- **Stress–strain graphs**

- Elastic hysteresis
- *Specified practical: Force-extension for rubber*
- Shape of F–x (or σ–ε) graph

- Tensile strain: $\varepsilon = \frac{\Delta l}{l}$
- E = gradient of σ–ε graph
- Tensile stress: $\sigma = \frac{F}{A}$
- Young modulus $E = \frac{\sigma}{\varepsilon}$
- *Specified practical: Young modulus of a wire*

- Strengthening
- Thin fibres
- Brittle fracture
- **Amorphous**
- **Brittle materials**
- Hooke's law up to fracture
- Elastic strain up to fracture

- Pre-compressed surface
- Surface imperfections
- Crack propagation

Q1 Hooke's law for a spring can be expressed by the equation:

$$F = kx$$

(a) State the meaning of the symbols, F, k and x, in this equation. [1]

F ....

k ....

x ....

(b) Express the unit of k in terms of the base SI units. [2]

....

....

....

(c) State the condition for this equation to be valid. [1]

....

Q2 Solid materials can be classified as *crystalline*, *amorphous* or *polymeric*. State what each of these terms means and give an example of each type of material. [3]

....

....

....

....

....

....

Q3

(a) Use the axes to sketch stress–strain (σ–ε) graphs for a brittle material, such as glass, and a ductile material, such as copper. Label the graphs. [2]

(b) Brittle materials are weak under tension. Describe one way of increasing the tensile strength of a brittle material and explain briefly how it works. [3]

....

....

....

....

....

Q4 In an experiment to determine a spring constant, Aled measured the extension of a spring when a load of mass (300 ± 6) g was hung from it. His result was (15.1 ± 0.2) cm.

(a) Calculate the value of the spring constant together with its **absolute** uncertainty. [4]

....

....

....

....

....

....

....

(b) A year-13 student told Aled that the period, T, of vertical oscillation of a mass, m, on a spring is related to the spring constant, k, by the equation:

$$T = 2\pi\sqrt{\frac{m}{k}}$$

Design a way in which Aled could use the result of his experiment in part (a) and another measurement to test this equation. [3]

....

....

....

....

....

....

Q5 Some materials are *ductile*.

(a) State what is meant by a ductile material. [1]

....

....

(b) Ductile metals are *polycrystalline*.

(i) State what polycrystalline means. [1]

....

....

(ii) Explain how the presence of dislocations accounts for the ductile nature of polycrystalline metals. A diagram may help your explanation. [3]

....

....

....

....

....

Q6 Ductile metals show *elastic strain* at low values of *stress*. Above the *elastic limit*, they exhibit *plastic strain*. State the meaning of the italicised terms. [4]

....

....

....

....

....

Q7 Wire **A** has length l and diameter d and is made from a metal with Young modulus E. It is joined at the end to wire **B** which has length $2l$, diameter $2d$ and Young modulus $1.5E$. A force, F, is applied to the end of each wire as shown.

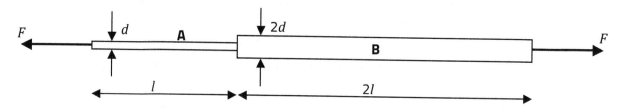

In the question that follows, σ_A is the strain in wire **A**, W_A is the work done in in extending wire **A**, etc. Complete the table to show the ratios of the given quantities. [6]

Quantity	Ratio
Tension	$\dfrac{F_A}{F_B} =$
Stress	$\dfrac{\sigma_A}{\sigma_B} =$

Quantity	Ratio
Strain	$\dfrac{\varepsilon_A}{\varepsilon_B} =$
Extension	$\dfrac{\Delta l_A}{\Delta l_B} =$

Quantity	Ratio
Work	$\dfrac{W_A}{W_B} =$
Potential energy per unit volume	$\dfrac{W_A/V_A}{W_B/V_B} =$

Space for calculations:

Q8 Joel used a piece of steel wire, of length 3.550 m, to obtain a value for the Young modulus, E, of steel. He put the wire under tension by hanging a load from its end and measured the diameter at various places. His results were:

Diameter /mm: 0.23   0.23   0.25   0.23   0.24   0.25

He added an extra load of 0.800 kg to the wire and measured the extra extension to be 3.1 mm.

(a) Why did Joel hang a load from the end of the wire before measuring the diameter?   [1]

....

(b) Estimate the **percentage** uncertainty in Joel's value for E, explaining your reasoning. There is no need to calculate E from Joel's results.   [3]

....

....

....

....

....

(c) Bethan said that using a thinner piece of wire would reduce the uncertainty in the result. Discuss whether she was correct.   [2]

....

....

....

....

Q9 A steel cable has a diameter of 3.0 cm and a length of 5.0 km. The steel has a *yield stress* of 300 MPa and a Young modulus of 2.0 GPa.

(a) State the meaning of the term in italics.   [1]

....

....

(b) In use, the maximum safe working stress is one fifth of the yield stress. Calculate the elastic potential energy stored in the cable with this stress.   [3]

....

....

....

....

....

....

Q10 The string of a modern longbow, used in archery contests, applies a maximum force of 280 N to a 50 g arrow, for a draw length (i.e. the distance the arrow is pulled backwards) of 76 cm.

(a) Assuming that the force is proportional to the distance the arrow is drawn back, calculate the work done in drawing the bow. [2]

....

....

....

(b) The efficiency of energy transfer to the arrow has been estimated as 90%. A sales brochure states that the bow is capable of shooting an arrow a distance of 400 m, if the arrow is shot at an angle of 45° to the horizontal. Evaluate this claim. [You should neglect the effects of air resistance.] [5]

....

....

....

....

....

....

....

Q11 It is proposed to place a set of energy-absorbing buffers at the end of a local railway line, to stop slow-moving trains from overshooting the railway platform.

The buffers contain springs which are compressed and absorb the energy of the trains. Two designs are proposed with different spring constants. Evaluate the advantages and disadvantages of using a spring with a lower spring constant. [4]

....

....

....

....

....

....

....

Question and mock answer analysis

Q&A 1 A student investigates the tensile properties of a rubber band of original length 10 cm. She produces this graph:

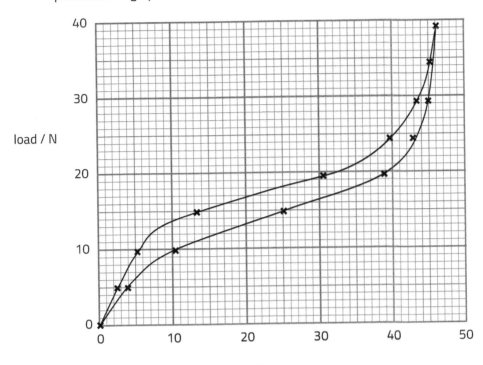

(a) Use the graph to help explain the terms *elastic* and *hysteresis*. [3]

(b) The crosses represent the results obtained by the student. Suggest briefly how the student carried out the investigation, including how the values of force were obtained. [3]

(c) Another student says that the results are not satisfactory between loads of 5 N and 15 N. Explain this comment and suggest an improvement to the method. [2]

(d) Rubber is different from metals in that large strains are possible and low tensile forces are involved. Account for these differences in terms of molecular structure. [4]

(e) Estimate the energy dissipated in loading and unloading the rubber band, explaining how you obtained your answer. [3]

What is being asked

This is a question centred on a specified practical. It has aspects of experimental design – parts (b) and (c) – the recall of an explanation of the mechanical properties of rubber – part (d) – and interaction with the data – parts (a), (b) and (e).

Mark scheme

Question part			Description	AOs			Total	Skills	
				1	2	3		M	P
(a)			An elastic material returns to its original size and shape when a stress is removed. [1] <u>In this case there is zero extension at the end of the unloading curve.</u>	1			3		
			In elastic hysteresis, load–extension graphs for unloading and loading do not coincide. [1] <u>In this case the unloading curve is underneath the loading curve.</u>	1					
			Both underlined sections. [1]		1				

Question part			Description	AOs			Total	Skills	
				1	2	3		M	P
(b)			The student had a set of loads (or masses) which she added to the rubber band up to the maximum, measuring the extension each time. [1]	2	1		3	1	3
			She repeated the measurements whilst decreasing the load (in the same way). [1]						
			Numerically relating weight to masses, e.g. 3.8 N is produced by a mass of 4.0 kg. [1]						
(c)			On the loading curve, there is a marked change of slope between 5 N and 15 N (so the shape of the curve is uncertain). [1]			2	2		2
			Obtain results for different masses (e.g. 0.7, 0.9, 1.2 kg) between. [1]						
(d)			Extensions in other materials, e.g. steel, are produced by stretching very stiff [inter-atomic] bonds. [1] These bonds are short range and so the material breaks at small strains [even if the material is ductile]. [1]	4			4		
			In rubber, large extensions are produced by the straightening out of long C–C chains [1] [which is possible by the rotation of the single bonds], and needs much lower forces. [1]						
(e)			Identification of the area between the curves as the dissipation. [1]	1			3		1
			Reasonable method attempted, e.g. counting squares or division into geometrical shapes. [1]		1				
			Answer in range 1.3–1.7 J [1]		1				
Total				9	4	2	15	2	5

Rhodri's answers

(a) Elastic materials go back to their original shape. ✗ (no bod)

Hysteresis is where there is a loop between the two graphs. ✓ (bod)

MARKER NOTE

Rhodri doesn't achieve the first marking point; he should have said that the original shape and size were regained when the stress was removed, or words to that effect. In the second, the expression 'loop between the graphs' is slightly unclear but effectively is the same as required and the examiner awards the mark, by bod.

No reference is made to the actual graph so the third marking point is not awarded.

1 mark

(b) The loads are multiples of just less than 5 N, which is the weight of a 0.5 kg weight. ✓ The student added 0.5 kg weights one by one and measured the extension, ✓ then removed the loads one after the other and measured the extension. ✓

MARKER NOTE

Rhodri has correctly observed that the increments of load were the weight of a 0.5 kg mass. It might have been nicer if he had used $W = mg$ to show this more explicitly but he has done enough to obtain the mark. He has proceeded to relate this to the standard method of this experiment for the other two marks.

3 marks

(c) He should repeat the readings and take an average to make sure they are more accurate (or check the accuracy). ✗

MARKER NOTE

Rhodri has failed to spot that this question was about the actual results obtained. He has given the weak answer of 'repeat and average' when this will not solve the problem in this case. More data are needed to fill in the gap and hence determine the precise shape of the graph here.

0 marks

(d) This rubber band had an extension of 46 cm for an original length of 10 cm and the load was only 38 N. If it had been steel, the extension would have been a lot less – just a few mm. This is because rubber is made of long-chain molecules, which can stretch out. ✓

MARKER NOTE

For most of his answer, Rhodri has just repeated the question, albeit with the addition of some numerical data. The only new aspect, for which he obtains credit, is the unravelling of tangled molecules. He has not explained why steel is stiff or why it can only stretch a short distance. He should also have explained why large extensions for rubber were achieved with low forces.

1 mark

(e) Average distance between graphs = 3 N (by eye)

Distance stretch = 46 cm = 0.46 N ✓

∴ Area between graphs = 0.46 × 3 = 1.38 J

Say 1.4 J ✓

MARKER NOTE

A nice way of estimating the 'area' between the curves (for a mark) and an answer well within the expected range (for another). Rhodri misses the easiest mark by not mentioning the significance of this area in terms of the dissipation of energy.

2 marks

Total **7 marks / 15**

Ffion's answers

(a) The graph shows that, when the load is removed, the extension goes back to zero. This is what is meant by elastic. ✓

The graph shows that the load-extension graphs for loading and unloading are different ✓ – this is elastic hysteresis. ✓

MARKER NOTE

This was a good answer. Ffion correctly explains both *elastic* and *hysteresis* and identifies the features of the graph which illustrate these properties.

A good concise answer.

3 marks

(b) The student had a set of weights up to 3.8 N. She put them on, one after the other, and measured the extension from the beginning each time. ✓ She then removed the weights one by one and measured the extensions (from the beginning) again. ✓

MARKER NOTE

Ffion has correctly described the procedure for loading and unloading, together with determining the extension for this specified practical, and hence obtained marks two and three. She identified the 3.8 N maximum load but doesn't relate this to the masses needed to achieve this weight and so misses the numerical part of the question.

2 marks

(c) There should really be more results. It's a complicated graph, so to be sure you need more readings. ✓ bod, ✗

MARKER NOTE

Ffion has something of the right idea. The examiner has awarded a mark bod. for *more data* because of the phrase 'complicated graph'. To obtain the second mark, Ffion should have gone on to explain the nature of the complications, i.e. large change of gradient, and the need for more data in the specified area.

1 mark

(d) Rubber is a tangle of long molecules. These can be straightened out a lot, so large extensions are possible. ✓ This doesn't take much force because bonds are not being stretched – just rotated. ✓

Metals are different because the bonds between the atoms need stretching which takes much bigger forces. ✓

MARKER NOTE

Ffion obtains both of the marks available for the properties of rubber – extensible and low tension needed – and obtains the marks.

Her answer as to why large forces are needed to stretch steel is correct and elicits a mark. She makes no attempt at an explanation of the fact that steel can only extend a few percent before fracture occurs, which is needed for the fourth mark.

3 marks

(e) Area under graph is the work done:

To find this, count the squares:

A 1 cm square = 0.05 m × 5 N = 0.25 J ✓

Stretching: Total number of squares = 31

∴ Work done = 7.75 J ✓

Contracting: 25 squares → 6.25 J

∴ Energy lost as heat = 7.75 − 6.25 = 1.5 J ✓

MARKER NOTE

Ffion quite reasonably estimates the work done (on the rubber band) in stretching the rubber band, by square counting. She also estimates the work done (by the band) in contracting. She recognises, and states, that the difference between these is the 'loss' in energy as 'heat', which is acceptable at AS and picks up all three marks.

3 marks

Total **12 marks / 15**

Q&A 2 Briefly describe the process by which a brittle material breaks under tension. [3]

What is being asked

This is a straightforward AO1 question which asks candidates to reproduce knowledge which is required by the specification. It is likely to be asked as part of a longer question.

Mark scheme

Description	AOs			Total	Skills	
	1	2	3		M	P
Cracks or imperfections in [surface of] material [1]						
Stress is concentrated at the tips of cracks, exceeding breaking stress [locally] [1]	3			3		
Cracks extend [further concentrating stress] [1]						
	3			3		

Rhodri's answers

Brittle materials break because of cracks or scratches on their surface. ✓ An example is builders 'cutting' bricks by making a crack across them and hitting them. ✗

MARKER NOTE
Rhodri has correctly stated that brittle fracture is initiated by cracks on their surface and gains the first mark. Instead of giving more details of the process, which would gain him more marks, he proceeds to give an example, which is not required.

1 mark / 3

Ffion's answers

Brittle materials, such as glass, have small holes or cracks in them. ✓ When the material is stretched, the stress at the edge of the cracks is much bigger than the average stress. If this stress is big enough, the bonds at the end of the crack break, ✓ which makes the crack grow until it extends across the material. ✓ So the material breaks.

MARKER NOTE
Ffion has given a good account of the process of brittle fracture. She has identified the significance of cracks for the first mark and correctly stated that the stress at the tip of a crack is magnified. She talks correctly of this stress breaking bonds, for which the examiner awards the second mark. Ideally she would have said that the stress at the new tip of the crack is now even greater, leading to runaway crack growth, but the marking scheme does not require this and she is awarded the third mark for the statement that the crack grows..

3 marks / 3

Section 6: Using radiation to investigate stars

Topic summary

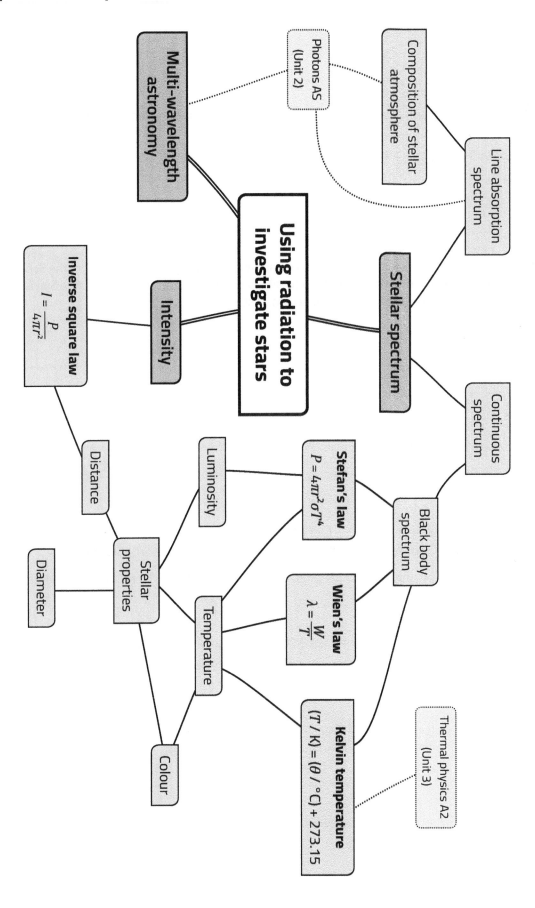

Q1 The joule (J) may be written in terms of the SI base units as kg m² s⁻². The SI unit for the *intensity* of radiation is W m⁻². Use defining equations to express this unit in SI base units. [3]

Q2 Define the term *black body* in terms of both the absorption and the emission of radiation. [2]

Q3 Red dwarf stars of the class M5V have radii approximately one quarter that of the Sun and a temperature (in kelvin) approximately half that of the Sun. The luminosity of the Sun is approximately 4×10^{26} W. Estimate the luminosity of M5V stars. [4]

Q4 The mass, M, and luminosity, L, of stars are often expressed in terms of the mass and luminosity of the Sun, $M_\odot$ and $L_\odot$ respectively. The greater the mass, the greater is its luminosity.

For masses in the range $0.43M_\odot < M < 2M_\odot$, the relationship in literature is often given as

$$\frac{L}{L_\odot} = \left(\frac{M}{M_\odot}\right)^4.$$

(a) The star 61 Cygni A has a mass of 0.70 $M_\odot$. Use the above equation to estimate its luminosity as a multiple of $L_\odot$. [1]

(b) In fact the luminosity of 61 Cygni A is 0.153 $L_\odot$. Alex says that the power of $(M/M_\odot)$ in the above equation should be less than 4. Evaluate whether he is right. [2]

Q5 The graph is of the continuous spectrum of the light from a distant star. The distance of the star from the Earth is known. The total intensity of the star's radiation arriving at the Earth is given by the area under the graph.

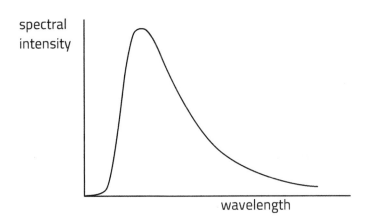

Explain how such a graph (with appropriate axis scales) can be used, together with the laws of radiation, to determine properties of the star. [6 QER]

....

....

....

....

....

....

....

....

....

....

....

....

....

Q6 The spectrum of a star consists of a *continuous emission spectrum* and a *line absorption spectrum*. Describe what is meant by these terms. [2]

....

....

....

....

Q7 Explain how information about a star's composition may be derived from its absorption spectrum. [3]

....

....

....

....

....

Q8 The diameter of the Sun is 1.39×10^6 km and the wavelength of the peak emission is 501 nm.

(a) Calculate the temperature of the solar surface. [2]

....

....

....

....

(b) A website gives the solar luminosity as 3.83×10^{26} W. Evaluate whether the data in this question are consistent with the Sun emitting radiation as a black body. [3]

....

....

....

....

....

Q9 The surface of the Sun has a temperature of about 6000 K. Sunspots are small regions of the solar surface which have a temperature of about 4000 K.

(a) (i) Calculate the peak wavelength of the radiation emitted by a sunspot. [2]

....

....

....

(ii) Identify the region of the e-m spectrum in which the peak wavelength lies. [1]

....

(b) Sunspots appear black in images of the solar surface. Explain this. A calculation will help your answer. [3]

....

....

....

....

....

Q10 Bryn says that, if Wien's law is correct, the photon energy of the peak of a star's spectrum is proportional to the temperature of the surface of the star. Evaluate whether he is correct. [3]

....

....

....

....

....

Q11 Describe what is meant by multiwavelength astronomy and give an example of its use. [2]

....

....

....

....

Q12 The table gives information about the temperature, T, of various sources of thermal radiation in the universe.

Source	T / K
Cosmic microwave background radiation	2.7
Galactic molecular clouds (from which stars form)	10 – 50
Red giant / Blue supergiant star surface	3 000 / 50 000

Source	T / K
Black hole inner accretion disk	10^7
Gas between galaxies (typical)	10^6
Supernova	5×10^7

Use data from this table to explain how multiwavelength astronomy is useful in studying the different processes which occur in the universe. [5]

....

....

....

....

....

....

....

....

....

....

....

Unit 1 Practice questions

Q13 The UV section of the e-m spectrum has a wavelength range of 10–400 nm. Calculate the range of photon energies in the UV spectrum. Express your answers in both J and eV. [4]

Q14 Singly ionised helium atoms, He⁺ (i.e. helium atoms with one electron missing) are found in the atmospheres of hot stars. The bottom 6 energy levels in He⁺ are:

–54.4 eV (ground state)   –13.6 eV   –6.0 eV   –3.4 eV   –2.2 eV   –1.5 eV

(a) An absorption line of wavelength 1.0 μm is observed in the infra-red spectrum of a star. Explain how this can arise from He⁺ in the atmosphere of the star. [4]

(b) The energy of photons in the visible spectrum lies between 1.9 eV and 3.1 eV. Identify transitions in He⁺ ions which can lead to dark lines in the visible absorption spectrum. [2]

(c) Eleri notices that there are no lines due to He⁺ in the visible spectrum of the Sun. She claims that this is because the temperature of the Sun's surface is too low, about 6000 K. Evaluate this claim. [3]

Question and mock answer analysis

Q&A 1   To a good approximation, stars behave as *black bodies*.

(a) State what is meant by a black body in terms of both the emission and absorption of radiation.   [2]

(b) State what is meant by the *luminosity* of a star.   [1]

(c) Three of the characteristics of the Sun are as follows

    Diameter = 1.4×10^6 km   Surface temperature = 5770 K   Luminosity = 3.83×10^{26} W

In the future, the Sun will pass through a stage in which its diameter is 300 million km and its surface temperature is 3000 K. It will then become a much smaller star with a diameter of 14 000 km and surface temperature 20 000 K.

Use the above information, together with calculations, to describe the changes in the appearance of the Sun to a distant observer.   [5]

What is being asked

Like many questions, this one starts with some aspects of recall (AO1), designed to lead you into an application of physics (AO2). In this case, you should have learned the definitions of *black body* and *luminosity* to be able to answer parts (a) and (b). In answering part (c), you will need to identify the observational features which are amenable to calculation, using the Stefan-Boltzmann's law and Wien's law, do the relevant calculations and then interpret the results.

Mark scheme

Question part		Description	AOs			Total	Skills	
			1	**2**	**3**		**M**	**P**
(a)		[A black body is one which] absorbs all [electromagnetic] radiation which falls upon it … [1] …[and it] emits the most radiation possible at any wavelength [and that temperature]. [1]	2			2		
(b)		The luminosity of a star is the power emitted [or the energy emitted per unit time] in the form of electromagnetic radiation. [1]	1			1		
(c)		Use of Stefan's law attempted [even with slips], e.g. substitution in $L = A\sigma T^4$ (accept radius/diameter confusion or πr^2 used) at any stage or $\frac{L_1}{L_2} = \frac{A_1 T_1^4}{A_2 T_2^4}$ seen. [1] Use of Wien's law attempted [even with slips] [1] Luminosity (or $L / L_\odot$) **and** peak wavelength of either evolved stage calculated [1.3×10^{30} W / $3000\,L_\odot$ 970 nm; 5.6×10^{24} W / $1.5 \times 10^{-2}\,L_\odot$ 150 nm] [1] [or luminosity or peak wavelength of both] Characteristics of both stages calculated. [1] Identification of colours [red $\longrightarrow$ blue/white] or region of e-m spectrum of peaks [IR, UV] [1]		5		5	4	
Total			3	5		8	4	

Rhodri's answers

(a) A black body absorbs and emits all radiation which falls on it. ✓ ✗

MARKER NOTE

Some confusion in Rhodri's answer. A black body absorbs all radiation which is incident upon it. Emitting 'all radiation which falls on it' makes no sense. It emits radiation more strongly at any wavelength than a non-black body. **1 mark**

(b) This is the total power output ✗ Not enough

MARKER NOTE

Rhodri was almost there but he needed to specify electromagnetic radiation. Stars also give out energy in the form of neutrinos, which does not contribute to the luminosity. **0 marks**

(c) The next stage is a Red giant – it would appear red and much brighter (more luminosity). The last stage is a White dwarf and much fainter. ✓

Red giant: $L = A\sigma T^4$

$= 4\pi \, (3.0 \times 10^8)^2 \times 5.67 \times 10^{-8} \times 3000^4$ ✓ attempt

$= 5.19 \times 10^{24}$ W

$\lambda_{max} = \dfrac{2.90 \times 10^{-3}}{3000} = 9.67 \times 10^{-7}$ – so red ✓

White dwarf: $L = A\sigma T^4$

$= 4\pi \, (1.4 \times 10^4)^2 \times 5.67 \times 10^{-8} \times 20000^4$ ✗

$= 2.2 \times 10^{19}$ W – much fainter

$\lambda_{max} = \dfrac{2.90 \times 10^{-3}}{20000} = 1.45 \times 10^{-7}$ – so white ✓

MARKER NOTE

The first mark that Rhodri achieves is the last one in the mark scheme: he correctly identifies the colours of the star in the two stages.

He picks up one mark each for using the Stefan-Boltzmann law and the Wien displacement law. He makes the same mistakes in the calculation of L for both stars: he fails to convert km to m and he uses the diameter instead of the radius in calculating the surface area. However, he correctly calculates the peak wavelength of both so he gains the 4th mark. The absence of the unit for λ_{max} is not further penalised. **4 marks**

Total **5 marks / 8**

Ffion's answers

(a) It absorbs all radiation which falls on it. ✓

It emits all wavelengths of radiation better than any other kind of body ✓

MARKER NOTE

The first part of the answer is good. Ideally, the second part of the answer would have mentioned that the comparison object was at the same temperature, e.g. a black body at 1000 K emits more radiation at all wavelengths than does a non-black body at 1000 K. However, this omission was not penalised on this occasion. **2 marks**

(b) Luminosity is the energy of the electromagnetic radiation given out every second. ✓

MARKER NOTE

A short and entirely correct answer. A well-learned definition. **1 mark**

(c) Main sequence $\rightarrow$ Red giant $\rightarrow$ White dwarf ✓

Luminosity: First stage,

$L = A\sigma T^4 = 4\pi (1.5 \times 10^{11})^2 \times 5.67 \times 10^{-8} \times (3000)^4$ ✓

$= 1.30 \times 10^{30}$ W $= 3400 \times$ L for Sun

Luminosity: Second stage

$L = A\sigma T^4 = \pi (1.4 \times 10^7)^2 \times 5.67 \times 10^{-8} \times (20000)^4$

$= 5.58 \times 10^{24}$ W $= 0.015 \times$ L for Sun ✓

Colour: Red giant

Peak $\lambda = \dfrac{W}{T} = \dfrac{2.90 \times 10^{-3}}{3000} = 9.7 \times 10^{-7}$ m ✓

This is in the infra-red region so the star will look red (and bright) as expected.

Colour: White dwarf

Peak $\lambda = \dfrac{W}{T} = \dfrac{2.90 \times 10^{-3}}{20000} = 1.5 \times 10^{-7}$ m ✓

This is in the UV so the star will appear bluish white and faint as expected.

MARKER NOTE

Ffion also achieves the 5th mark from the first part of her answer – a recollection of GCSE learning! Really, this answer should come out of the calculations but it scores the marks anyway.

Ffion uses both the Stefan-Boltzmann and Wien's laws. It appears she has made a slip in calculation of the luminosity of the white dwarf: she uses πd^2 instead of $4\pi r^2$. These, of course, give the same answer and the examiner awards the mark bad.

Her Wien calculations are correct. Hence she scores all 4 calculation marks, and her final written descriptions would have earned her the 5th mark if she hadn't already achieved it! **5 marks**

Total **8 marks / 8**

Section 7: Particles and nuclear structure

Topic summary

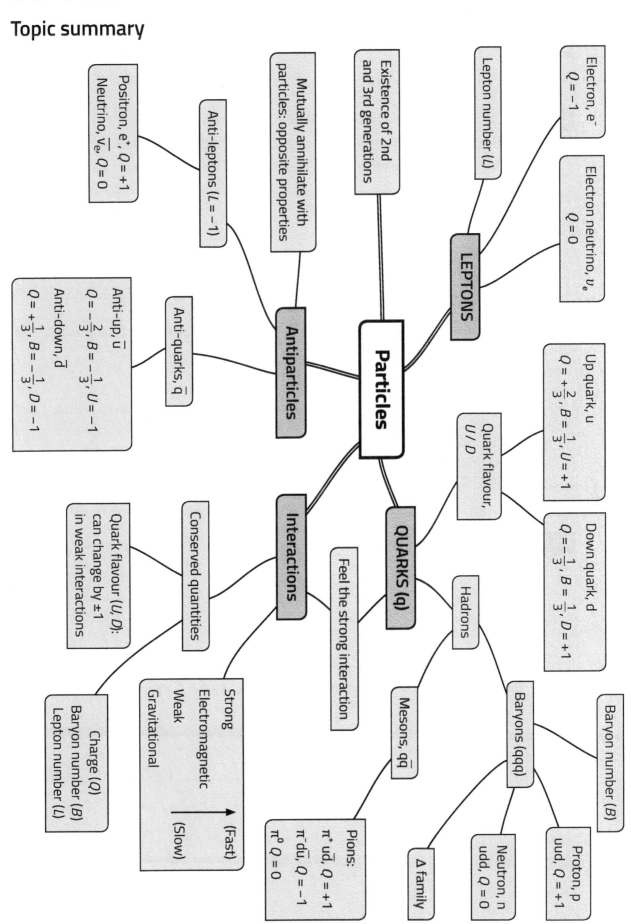

Q1 Electrons can be described as *fundamental particles*. Protons are *composite particles*. Explain this difference for these particles. [2]

....

....

Q2 Particle X can interact through the strong interaction. What can you conclude about X? [1]

....

Q3 The following is a selection of first-generation particles and antiparticles:

electron (e^-)　proton (p)　anti-neutrino ($\overline{v_e}$)　pi+ meson (π^+)　anti-neutron ($\overline{n}$)　positron (e^+)

Identify which of these particles can interact via each of the strong, electromagnetic and weak forces. [2]

Strong: ....

Electromagnetic: ....

Weak: ....

Q4 Neutral pions (π^0) consist of a mixture of $u\overline{u}$ and $d\overline{d}$. They decay into two photons with a lifetime of about 10^{-16} s.

(a) State which interaction is responsible for this decay. [1]

....

(b) Give two reasons for your answer. [2]

(i) ....

(ii) ....

Q5 (a) Name and write the symbol for the anti-particle of each of the following particles: [3]

(i) e^- ....

(ii) n ....

(iii) v_e ....

(b) Eurig says that π^- is the anti-particle of π^+. Determine whether Eurig is correct. [2]

....

....

....

....

Q6 The Δ^{++} is a first-generation particle.

(a) Keira says that Δ^{++} must be a baryon. Explain why she is correct and give its quark structure. [2]

....

....

....

(b) The Δ^{++} has a lifetime of about 10^{-24} s. It decays into a proton with the emission of another first-generation particle.

(i) State the interaction which is responsible for the decay and give a reason for your answer. [2]

....

....

(ii) Momentum and energy are conserved in this decay. State the other quantities which are conserved in this sub-atomic particle decay. [2]

....

....

(iii) Write the equation for this decay and show how each of the quantities you named in (ii) is conserved. [3]

....

....

....

....

....

Q7 Sub-atomic particles can be classified as *leptons*, *baryons* and *mesons*. The positive pion, π^+, decays into a positron, e^+, and a neutrino, ν_e.

(a) Classify the particles, π^+, e^+ and ν_e. [2]

....

....

....

....

(b) Explain which properties of particles are conserved in this decay, in addition to momentum and energy. [3]

....

....

....

....

(c) State a particle property which is not conserved in this decay and explain your answer briefly. [1]

....

....

Q8 Two protons collide at high energy. The following reaction cannot occur.

$$p + p \rightarrow \Delta^+ + e^- + \pi^+$$

where Δ^+ is a first-generation baryon.

(a) Explain which of the conservation laws this reaction would violate. [3]

....

....

....

....

....

(b) Usually a Δ^+ decays into a nucleon and a pion, e.g. $\Delta^+ \rightarrow p + \pi^0$, with a lifetime of $\sim 10^{-24}$ s. Rarely, however, the following decay is observed:

$$\Delta^+ \rightarrow p + \gamma$$

Explain why this is a much slower decay. [2]

....

....

....

Q9 A neutron and a positive pion collide at high energy. A student suggests that the following reaction might be observed:

$$n + \pi^+ \rightarrow \Delta^{++} + e^-,$$

where the Δ^{++} is a first-generation baryon.

Explain whether this reaction would violate the conservation of lepton number, charge and baryon number. [3]

....

....

....

....

....

Q10 Quarks are said to have a *baryon number* of $\frac{1}{3}$. Explain, using examples, how this is consistent with each of the following:

(a) Baryons each consist of three quarks. [2]

....

....

....

(b) Mesons each have a baryon number of 0 (zero). [2]

....

....

....

Q11 The Sun emits the Solar Wind, which is a stream of particles, mainly electrons and protons. The nuclear reactions in its core also produce a stream of neutrinos. If these particles were to hit a piece of lead, whilst the electrons and protons would have a range of less than 1 cm, the range of the neutrinos has been estimated as 1 light year.

Explain this difference by considering the kinds of interaction in which the three particles can engage.

[4]

Q12 The laws of conservation of energy and momentum apply to all collisions. Write a brief account of the **additional** conservation laws which apply to interactions between sub-atomic particles.   [6QER]

Q13 When an isolated particle decays, the conservation of mass and energy requires that the total mass of the resulting particles is lower than the mass of the decaying particle.

The following table contains the masses of **all** first-generation particles as multiples of the electronic mass, m_e.

Particle	Symbol(s)	Type	Mass / m_e
neutrino	ν	lepton	$<10^{-5}$
electron	e^-	lepton	1
pion	$\pi^+ / \pi^0 / \pi^-$	meson	207
proton	p	baryon	1836
neutron	n	baryon	1839
delta	$\Delta^{++} / \Delta^+ / \Delta^0 / \Delta^-$	baryon	2411
rho	$\rho^+ / \rho^0 / \rho^-$	meson	1516

Second- and third-generation particles of each type have greater masses.

Use this information to explain each of the following statements. For some of the statements you will need to consider other conservation laws.

(a) The Δ^+ particle can decay into a proton and a π^0 but not a proton and a ρ^0. [2]

....

....

....

....

(b) The π^0 can decay into an electron, a positron and a photon. [2]

....

....

....

....

(c) The neutron can only decay by means of the weak interaction. [3]

....

....

....

....

(d) The proton is a stable particle, i.e. it does not decay. [2]

....

....

....

....

Question and mock answer analysis

Q&A 1 Particle Y decays, producing a neutral pion, π^0, a positron, e^+, and a lepton, Z, with no charge.

(a) Imran says that particle Y could be a positive baryon. Josephine thinks that it could be a neutral meson. Evaluate which, if either, is correct. [4]

(b) Jemima says that Z must be a neutrino. Evaluate this statement. [2]

(c) Explain which interaction, strong, weak or electromagnetic, is responsible for the decay of particle Y. [2]

(d) Explain how a measurement of the mean lifetime of Y particles would provide supporting evidence for your answer to (c). [2]

What is being asked

This is quite a tricky question. Parts (a) and (b) require candidates to use conservation laws in particle physics to come to conclusions about the nature of particles. Because the question does not mention conservation laws, these two parts of the question are classed as AO3. Parts (c) and (d) test the candidates' knowledge of the different kinds of interactions that sub-atomic particles undergo. The basis of the analysis is given in part (c) so this is an AO2 question; in part (d) the candidates are pointed very much in the direction of recalled properties of the weak interaction and so it is AO1.

Mark scheme

Question part	Description	AOs 1	AOs 2	AOs 3	Total	Skills M	Skills P
(a)	Statement of any conservation law (charge, baryon number or lepton number) [1] e.g. 'charge is conserved' Correct application to the decay [1] Correct application of a second law [1] Identification of particle, following correct reasoning, and evaluation [1]			4	4		
(b)	The only leptons with no charge are neutrinos [or anti-neutrinos] [1] Z cannot be an antineutrino because total lepton number would be 2 (which Y cannot have) so Jemima is correct [1]			2	2		
(c)	The reaction involves a neutrino [1] [Neutrinos only interact via the weak interaction] **hence** weak [1]		2		2		
(d)	Weak decays take longer than strong or e-m [1] If decay time longer than 10^{-10} s then weak [1]	2			2		
Total		2	2	6	10		

Rhodri's answers

(a) Y must have a positive charge because charge is always conserved ✓ and the product particles have charges of 0 + 1 + 0 = 1. ✓ So Josephine cannot be correct but Imran could be right.

There is a meson on the right so there must be a meson on the left ✗ so the particle is a positive meson (not clear ✗)

MARKER NOTE

Rhodri handles the conservation of charge well and correctly indicates that Y must have a positive charge. He thus obtains the first two marks on the scheme. He mistakenly thinks that mesons are conserved; so that, although his conclusion is correct it is arrived at by faulty reasoning.

2 marks

(b) A neutral lepton must be a neutrino ✓

so Jemima is correct (not enough ✗)

MARKER NOTE

The marking scheme allows an easily accessible first mark which Rhodri achieves. However, he does not rule out an anti-neutrino and so was not given the second mark.

1 mark

(c) The reaction involves a neutrino ✓, so it must be weak ✓ (just)

MARKER NOTE

The marking scheme generously allowed Rhodri's answer – the section in brackets in the scheme was not required.

2 marks

(d) Weak interactions are less likely to happen than strong. They have shorter range than electromagnetic interactions – so this reaction is unlikely to happen. ✗

MARKER NOTE

Rhodri's analysis relates to a collision interaction rather than to a decay. Hence he talks about the probability of a reaction occurring. This doesn't attract either point of the mark scheme.

0 marks

Total	5 marks /10

Ffion's answers

(a) Y cannot be a lepton because there are two leptons on the right. ✓

Y cannot be a baryon because there are no baryons on the right. ✓

Y must have a positive charge because the total charge on the right is +1. ✓

So the particle is a positive meson. (not clear ✗)

MARKER NOTE

Ffion doesn't actually state any conservation law explicitly but applies all three correctly. So the examiner has awarded the first mark slightly generously by implication. On the other hand, Ffion has not actually answered the question fully; to be awarded the last mark she needed to say to what extent each person was correct..

3 marks

(b) The lepton number of e^+ is –1, so the lepton number of Z is +1 so it must be either an electron or a neutrino ✓

Z is neutral so it must be a neutrino. ✓

MARKER NOTE

Ffion sets out her answer differently from the mark scheme but clearly addresses both marks correctly.

2 marks

(c) Leptons don't feel the strong force (because they have no quarks). Neutrinos are neutral so they don't feel the e-m force so it must be weak. ✓✓

MARKER NOTE

This answer by Ffion is much better than Rhodri's but she cannot get more than 2 marks. Again, her setting out was very different from that in the mark scheme.

2 marks

(d) Strong decays happen very quickly (about 10^{-24} s). Electromagnetic decays take longer (about 10^{-16} s). Weak decays take the longest ✓ ✗

MARKER NOTE

A very good start to the answer. Unfortunately she only achieves the first marking point – with her last statement. Unaccountably she misses out an estimate of the time for a weak decay.

1 mark

Total	8 marks /10

Q&A 2 Particles which contain quarks or anti-quarks are called hadrons. Sub-families of the hadrons are *baryons* and *mesons*.

(a) Compare baryons and mesons in terms of quark make-up and a conservation law. [3]

(b) Two high-energy protons collide and undergo the following reaction:

$$p + p \rightarrow p + X + \pi^+$$

where X is a first-generation particle.
Use conservation laws to identify particle X. Give your reasoning. [4]

Unit 1 Practice questions

What is being asked

This is intended to be a straightforward question. Part (a) requires you to remember details of the nature and properties of baryons and mesons. This is bookwork and hence AO1. Part (b) asks you to apply your knowledge to a given reaction. The mark scheme allows this to be done in terms of baryons or quarks. Notice that a simple identification of X will not score, even if correct, because the question asks for your reasoning. The basis of the analysis is given, so this is an AO2 question.

Mark scheme

Question part			Description	AOs			Total	Skills	
				1	**2**	**3**		**M**	**P**
(a)			Baryons are composed of 3 quarks [1]	3			3		
			Mesons are composed of a quark and an anti-quark [1]						
			[In any interaction] the baryon number [or number of baryons] is conserved but the number of mesons is not [1] **Both needed**						
(b)			Protons are baryons and π^+ is a meson [1]		4		4		
			Baryon number = 2, so X is a baryon [1]						
			Charge is conserved so X is neutral / uncharged [1]						
			Uncharged [first-generation] baryon is neutron (accept Δ^0) [1]						
			Alternative (in terms of quarks)						
			proton composition = uud; $\pi^+ = u\bar{d}$ ($\checkmark$)						
			To conserve charge, d must be created alongside $\bar{d}$ ($\checkmark$) (Accept: quark flavour is conserved)						
			Quarks on both sides = 4u + 2d ($\checkmark$)						
			$\therefore$ Composition of X = udd hence neutron ($\checkmark$)						
Total				3	4		7		

Rhodri's answers

(a) Baryons are uud or udd (where u is an up-quark and d is a down-quark). ✗

Mesons only have two quarks – one of which is an anti-quark, e.g. u$\bar{\text{d}}$ ✓ [bod]

Baryon number is conserved but mesons have no conservation laws. ✓ [bod]

MARKER NOTE

Rhodri answers in terms of specific baryons without naming them, rather than a general answer, which is not credited. His meson answer is also not ideal, not making it clear that it is a quark / anti-quark pair, but the examiner gives him the benefit of the doubt. He clearly states that baryons are conserved; mesons do obey the law of conservation of charge but his answer is close enough for the examiner to award the last mark.

2 marks

(b) Protons are uud quarks

To conserve quarks X + π⁺ must be uud. ✓

π⁺ is u$\bar{\text{d}}$ ✓, so the quark composition of X is udd, ✓ which is neutral because

$u = +\frac{2}{3}$ and $d = -\frac{1}{3}$. ✗ not enough

MARKER NOTE

Rhodri goes down the quark route and does rather well. The first mark he obtains is actually the second one on the mark scheme. The next mark is for both the proton and pion structures. He applies the conservation of quark flavour, although this is not clearly expressed for the third mark. He just misses out on the last mark because he doesn't actually say that X is a neutron.

3 marks

Total **5 marks /7**

Ffion's answers

(a) Baryons have 3 quarks, e.g. proton = uud ✓

Mesons have a quark and an antiquark, e.g. π⁺ = u$\bar{\text{d}}$ ✓

In collisions or decays, the baryon number is conserved ✗ [not enough]

MARKER NOTE

Ffion gives a textbook answer for the structure of both baryons and mesons. Her examples, the proton and positive pion, are not required for the first two marks.

She correctly mentions the conservation of baryon number. To nail the final mark she needed to say that there is no law of conservation of mesons.

2 marks

(b) In this reaction the protons are baryons, so the number of baryons must stay at 2. So X must be a baryon. ✗ [not enough] ✓

The meson has taken the positive charge ✓ [bod] so X must be neutral.

So X is a neutron. ✓

MARKER NOTE

Ffion clearly identifies protons as baryons. In order to obtain both the first two marks, she should have said that the π⁺ was a pion, so is not counted in the baryon number, leaving X to be a baryon.

She applies the conservation of charge; this could have been done more clearly, hence the bod comment. She correctly finishes off with identifying the neutral baryon as a neutron.

3 marks

Total **5 marks /7**

Unit 2: Electricity and Light

Section 1: The conduction of electricity

Topic summary

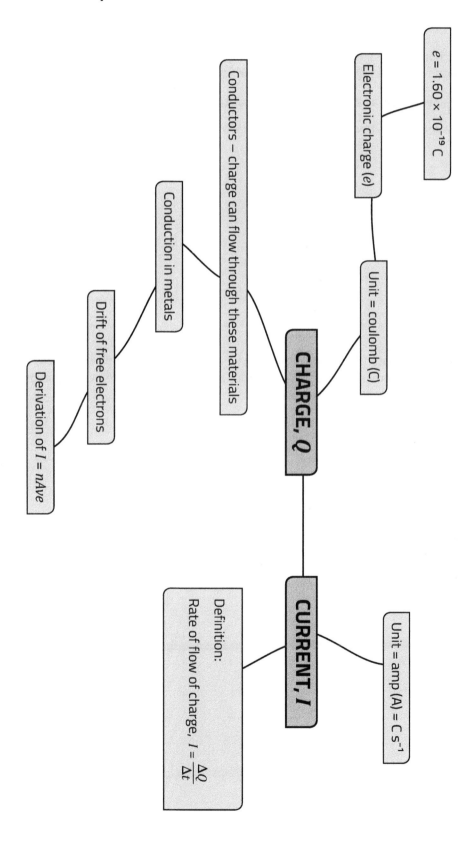

Q1 A light-emitting diode (LED) conducts a current of 15 mA. Calculate the number of electrons which flow into (or out of) the LED per minute. [e = 1.60 × 10⁻¹⁹ C] [2]

....

....

....

Q2 State what is meant by an electrical conductor. [1]

....

Q3 A capacitor is a device which stores electrical potential energy. It does this by holding equal positive and negative charges, Q, apart. The energy, W, stored is related to Q by the equation:

$W = \dfrac{Q^2}{2C}$ where C is a constant called the capacitance. The unit of C is the farad (F).

Express the farad in terms of the base SI units, m, kg, s and A. [4]

....

....

....

....

....

....

Q4 The radioactive material Am−241 is an alpha (α) emitter. A small sample of Am−241 has an activity of 37 kBq [that is, it emits 37 × 10³ alpha particles per second]. Calculate the magnitude of the electric current that represented by this activity. [2]

....

....

....

Q5 Wire A is made from a metal with 3.0 × 10²⁸ free electrons per m³. Its diameter is 0.60 mm and it carries an electric current of 1.5 mA. The values for wire B are 1.0 × 10²⁸ m⁻³, 0.30 mm and 10 mA. Calculate the ratio v_A/v_B, where v_A and v_B are the drift velocities of the free electrons in wire A and wire B respectively. [3]

....

....

....

....

....

....

Q6   A light-dependent resistor (LDR) is an electronic device made from a material in which there are no free electrons in the dark. However, photons of light can temporarily eject electrons from atoms, so that they can move through the material.

Nigel and Iestyn decide to investigate the current, I, in an LDR when it is illuminated by a light source at different distances, d.

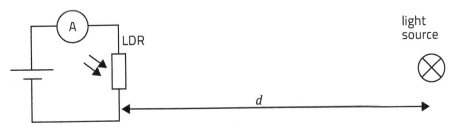

(a)   Nigel says that he expects the current to be inversely proportional to the square of the distance from the light source. Explain whether or not you agree with this statement.   [3]

....

....

....

....

(b)   The students' results are plotted below.

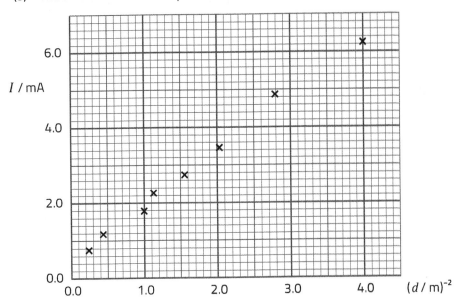

Evaluate to what extent the results agree with Nigel's hypothesis and suggest an explanation for any discrepancies.   [4]

....

....

....

....

....

....

....

Unit 2 Practice questions

Question and mock answer analysis

Q&A 1 Since May 2019, the charge on the proton has been defined as exactly $1.602\,176\,634 \times 10^{-19}$ C. The electronic charge is equal in magnitude but negative.

(a) A wire carries a current of (1.000 ± 0.005) mA. Evaluate what this means in terms of electron flow. [4]

(b) The current, I, in a metal wire is given by the equation:

$$I = nAve.$$

(i) Derive the equation. You may find it useful to include a diagram in your answer. [4]

(ii) Calculate the drift velocity of the electrons in an aluminium wire of diameter 0.50 mm which carries a current of 0.30 A, given that each atom of aluminium contributes three conduction electrons. [4]

Data: Density of aluminium = 2 710 kg m⁻³; mean mass of an aluminium atom = 4.48×10^{-26} kg

What is being asked

There are not many questions that an examiner can ask which contain concepts only from within this very short topic. This one almost succeeds but it does require knowledge of density, which is from Topic 1.1, Basic Physics. Part (a) is a calculation based upon the relationship between current and charge flow, with the additional requirement to take the uncertainty of current into account. The requirement to bring together the current/charge equation, $Q = It$, and electronic flow rates makes this AO3. Part (b) starts with the standard derivation of $I = nAve$, which you are expected to know. This is followed by the application of $I = nAve$ to the case of an aluminium wire. The tricky aspects here are the correct application of SI multipliers and remembering to use the radius of the wire, rather than the diameter, in calculating the cross-sectional area. Combining the density and atomic mass to give the concentration of electrons is quite demanding.

Mark scheme

Question part		Description	AOs			Total	Skills	
			1	2	3		M	P
(a)		Current = charge flow per second [1] or by impl. $$\text{Electrons} / s = \frac{(1.000 \pm 0.005) \times 10^{-3}\ A}{1.602... \times 10^{-19}\ C}\ [1]$$ $(= 6.242 \times 10^{15})$ Uncertainty = 0.5%, so electron flow $(6.24 \pm 0.03) \times 10^{15}\ s^{-1}$ [1] ecf With sf as above. [Accept $(6.242 \pm 0.031) \times 10^{15}$] [1]			4	4	3	
(b)	(i)	Drift distance in time $t = vt$ [1] Volume of wire with this length = Avt [1] Number of electrons drifting past any cross-section per second $[= nAvt/t] = nAv$ [1] ∴ Current [= charge per second] = $nAve$ [1] Alternatively, considering a period of 1 second can score full marks.	4			4		
	(ii)	In 1 m³ $n = 3$ [1] $\times \dfrac{2710\ kg}{4.48 \times 10^{-26}\ kg}$ [1] $(= 1.81 \times 10^{29})$ (or impl)		2				
		$A = \pi \times (0.25 \times 10^{-3}\ m)^2\ (= 1.96 \times 10^{-7}\ m^2)$ [1] (or impl)	1				4	
		$v = 5.3 \times 10^{-5}\ m\ s^{-1}$ [1] ecf on A and n			1	4		
Total			5	3	4	12	7	

Rhodri's answers

(a) 1 A = 1 coulomb per second

so 1 mA = 10^{-3} C s^{-1} ✓

1 C = 1.602 176 634 × 10^{19} electrons

So 1 mA = 1.602 176...× 10^{16} electrons per second ✗

So 0.005 mA ⟶ 0.008 × 10^{16} electrons.

So there are between 1.610 and 1.594 × 10^{16} electrons
per second. ✓✓ ecf

MARKER NOTE

This answer is very good but Rhodri has made one
mistake. He has obtained the first mark, equating
current to rate of charge flow. His mistake is to say
that $1/1.6 × 10^{-19}$ is $1.6 × 10^{19}$ electrons, so he lost
a single mark (the second one). The other working is
not as on the mark scheme, but he gives an equivalent
answer: the range of electron flows consistent with
the information and so he obtains the last two marks
on the ecf principle.

3 marks

(b)(i) In a section of wire of length vt
     the number of electrons is nAvt ✓

∴ $Q = nAvte$

$I = \dfrac{Q}{t} = \dfrac{nAvet}{t}$ ✓ $= nAve$

MARKER NOTE

At first sight, Rhodri's answer looks good but he doesn't explain
what he is doing. What is the significance of the distance vt,
for example? He doesn't identify any of his symbols. A better
approach for him would have been to draw and label a diagram,
such as:

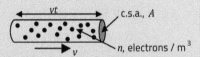

then use it to say
- all the electrons, numbering nAvt, within the volume will pass
  through the shaded area in time t.
- these electrons have charge, Q = nAvte
- and so the current, charge flow per second = nAve.

2 marks

(ii) $I = nAve = 0.30$ A

$v = \dfrac{I}{nAe}$

In 1 m³ the mass = 2710 kg

∴ $n = \dfrac{2710\,kg}{4.48 × 10^{-26}\,kg} = 6.05 × 10^{28}$ ✗ ✓

$A = π × (0.0025)2 = 1.96 × 10^{-5}$ ✗

So $v = \dfrac{0.30}{6.05 × 10^{28} × 1.96 × 10^{-5} × 1.60 × 10^{-19}}$

    $= 1.58 × 10^{-6}$ m s^{-1} ✓ecf

MARKER NOTE

Rhodri correctly works out the number of aluminium atoms per
m³ (the first ✓) but apparently identifies this with n, so doesn't
multiply by 3 (hence the first ✗). His method for calculating
A is correct but unfortunately, he doesn't convert from mm to
m correctly and he loses the A mark (the second ✗). However,
the mark for calculating the drift velocity is available and he is
awarded it on the ecf principle.

2 marks

Total   **7 marks /12**

Unit 2 Practice questions

Ffion's answers

(a) Current = charge per second. ✓

So 1.000 mA $= 1 \times 10^{-3}$ C s^{-1}

$1 C = \dfrac{1}{1.602\,176\,634 \times 10^{-19}}$ electrons

$\quad = 6.241\,509\,074 \times 10^{18}$ e/s ✓

$\therefore 1$ mA $= 6.241\,509\,074 \times 10^{15}$ e/s

If 1.005 mA, then 6.274×10^{15}

So Ans $= 6.242 \pm 0.03 \times 10^{15}$ e / s ✓ X

MARKER NOTE

Ffion's answer gains the first three marks. The first two are as in the mark scheme. Her method of determining the uncertainty in the number of electrons per second is to find the maximum flow rate, i.e. with a current of 1.005 mA. The uncertainty is then this maximum minus the value with $I = 1.000$ mA. To gain the last mark, she should have expressed the flow rate to the same decimal place as the uncertainty, as in the mark scheme.

3 marks

(b)(i) n = number of electrons per m^3

If the drift velocity is v all the electrons in a length vt will go past in time t. ✓

Volume of this length $= Avt$, ✓

so number of electrons $= nAvt$

So the number of electrons per second is nAv ✓

So the charge per second, which is the current, is $nAve$ ✓

MARKER NOTE

Ffion is able to communicate the logic of the derivation without using a diagram. She clearly identifies the meanings of n and v in the equation, and logically develops the sequence of ideas:

- the volume of the significant length of wire, vt.
- number of electrons passing a cross-section in a time, t
- the charge on these electrons
- hence the expression for the current.

4 marks

(ii) Atoms/m^3 $= \dfrac{2710 \text{ kg/m}^3}{4.48 \times 10^{-26} \text{ kg/atom}}$

$= 6.049 \times 10^{28}$ ✓

3 electrons per atom,

so $n = 1.815 \times 10^{29}$ per m^3 ✓

$A = \pi \times \left(\dfrac{5 \times 10^{-4}}{2}\right)^2 = 1.96 \times 10^{-7}$ m^2 ✓

$\therefore v = \dfrac{I}{nAe} = 1.76 \times 10^{-5}$ m s^{-1} X

MARKER NOTE

An almost perfect answer from Ffion but with an unfortunate slip at the end. She sets out her working clearly, calculating first the number density of the aluminium atoms (first mark) and then multiplying it by 3 (second mark) to give n. She correctly applies $A = \pi r^2$ for the third mark.

At the end she uses the incorrect value n, which appears to be a mistake but costs her a mark.

3 marks

Total   **10 marks /12**

Section 2: Resistance

Topic summary

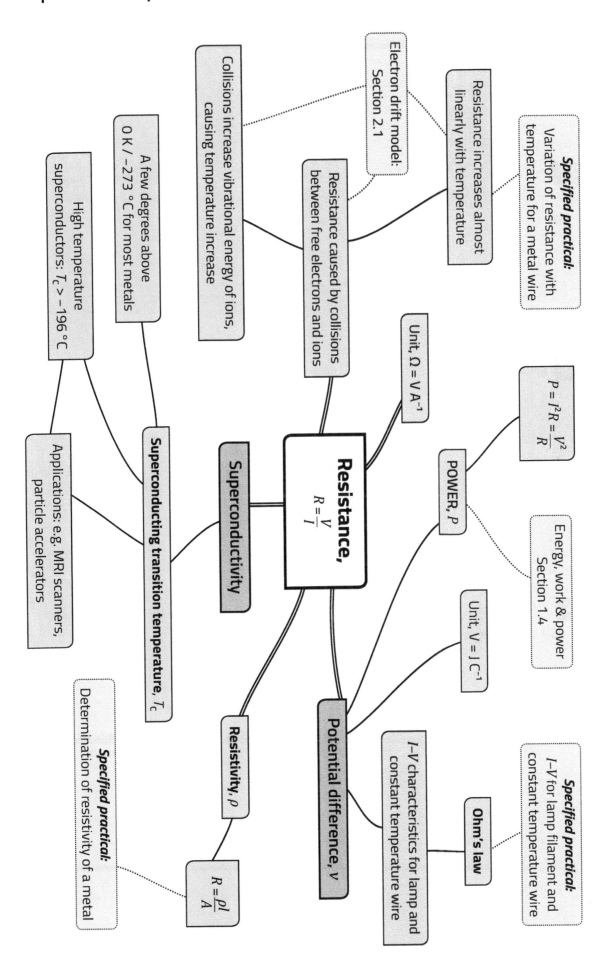

Electron drift model: Section 2.1

Resistance increases almost linearly with temperature

Specified practical: Variation of resistance with temperature for a metal wire

Collisions increase vibrational energy of ions, causing temperature increase

A few degrees above $0\ K / -273\ °C$ for most metals

High temperature superconductors: $T_c > -196\ °C$

Resistance caused by collisions between free electrons and ions

Applications: e.g. MRI scanners, particle accelerators

Superconducting transition temperature, T_c

Superconductivity

Resistance, $R = \dfrac{V}{I}$

Unit, $\Omega = V\,A^{-1}$

POWER, P

$P = I^2 R = \dfrac{V^2}{R}$

Energy, work & power Section 1.4

Unit, $V = J\,C^{-1}$

Resistivity, ρ

$R = \dfrac{\rho l}{A}$

Specified practical: Determination of resistivity of a metal

Potential difference, V

I–V characteristics for lamp and constant temperature wire

Ohm's law

Specified practical: I–V for lamp filament and constant temperature wire

Q1 A potential difference (pd) is applied across a component, X. As a result, there is a current of 1.5 A in X and 300 J of energy is transferred in a time 20 s.

Starting from definitions of potential difference and current, deduce the pd across X with clear reasoning. [3]

....

....

....

....

....

Q2 The joule (J), the SI unit of energy, can be expressed as $kg\ m^2\ s^{-2}$. Starting from definitions of pd and charge, express the volt (V) in terms of SI base units. [3]

....

....

....

....

....

Q3 (a) State Ohm's law. [1]

....

....

(b) Liam and Paul disagree over whether the equation $V = IR$ is a statement of Ohm's law. Explain to what extent it is. [2]

....

....

....

Q4 The current, I, in a wire can be related to the drift of free electrons by the equation $I = nAve$.

(a) Identify the symbols n, A, e and v in the equation. [2]

....

....

(b) A pd is applied across a wire and a current produced. If the temperature of the wire increases, the resistance of the wire increases. Account for this in terms of the electron drift model. [4]

....

....

....

....

....

....

Q5 Many metals exhibit superconductivity.

(a) State briefly what is meant by superconductivity. Include a graph and label the axes and significant features.   [3]

....

....

....

....

....

(b) State what is meant by high-temperature superconductors and explain their advantage in a named application.   [3]

....

....

....

....

....

....

Q6 A car filament lamp is designed to operate at 12 V. When the pd across it is 2.4 V, the current in the lamp is 1.5 A. When it is operating at the rated pd, the current is 3.0 A.

(a) Calculate the following ratios:

(i) $\dfrac{\text{Resistance of the filament at 12 V}}{\text{Resistance of the filament at 2.4 V}}$ [2]

....

....

....

(ii) $\dfrac{\text{Power of the lamp at 12 V}}{\text{Power of the lamp at 2.4 V}}$ [2]

....

....

....

(b) Explain briefly, in terms of a model of conduction, why the temperature of the filament increases with pd.   [2]

....

....

....

....

Q7 A reel of thin enamelled copper wire has a label with the following data:

0.1 mm   50 g   2.18 Ω/m   14 306 m/kg

(a) Select data from above and use it to estimate the resistivity of copper. [The 0.1 mm is the diameter.]

[3]

(b) A website gives the density of copper as 8.89 g cm⁻³. Evaluate whether this is in agreement with the above data.

[3]

Q8 Peter and Sion use a digital micrometer, a resistance meter and metre rule to determine the resistivity of constantan in the form of a wire. They obtained the following values:

diameter 0.32 ± 0.01 mm   length 2.000 ± 0.002 m   resistance 13.9 ± 0.1 Ω

(a) Use the data to determine a value for the resistivity along with its absolute uncertainty. Give your answer to an appropriate number of significant figures.

[4]

(b) They had another reel of constantan wire with a diameter of approximately twice that of their original wire. Peter said that they would get a value of resistivity with a lower uncertainty if they used this second reel because the percentage uncertainty in the diameter would be less. Sion said they'd be better just using a longer piece of wire.

Evaluate their suggestions.

[4]

Q9 A student finds an old filament-lamp labelled 240 V, 60 W. She uses a resistance meter to determine its room-temperature resistance. The result is 80 Ω. Estimate the temperature of the filament of the lamp in normal operation, assuming that the resistance of the filament is approximately proportional to its absolute temperature. [3]

....

....

....

....

....

Q10 The owner of a tropical fish tank wants to make a 10 W, 30 V electrical heater to maintain the water temperature. He decides to use a heater wire made from constantan, an alloy with an almost constant resistance over a wide range of temperatures.

(a) Explain an advantage of using a material with a constant resistance when making a heater filament. [2]

....

....

....

(b) The constantan wire has a diameter of 0.12 mm. Calculate the length of wire which the owner should use for the heater.
Resistivity of constantan = 4.9×10^{-7} Ω m. [4]

....

....

....

....

....

....

Q11 Wire **A** has diameter D, length l and is made from a material of resistivity ρ.

Wire **B** has a diameter $2D$, length $3l$ and is made from a material of resistivity 2.5ρ.

The two wires are separately connected to power supplies with the same pd, V.

Giving your reasoning, determine the ratio

$$\frac{\text{power dissipated in wire } \textbf{A}}{\text{power dissipated in wire } \textbf{B}}$$ [3]

....

....

....

....

....

Q12 (a) A coil of iron wire is contained in a test tube of oil with its ends projecting. Describe briefly a method for investigating the variation of the resistance of this wire over a range of temperatures between 0 °C and 100 °C. Sketch a graph of the expected results. [4]

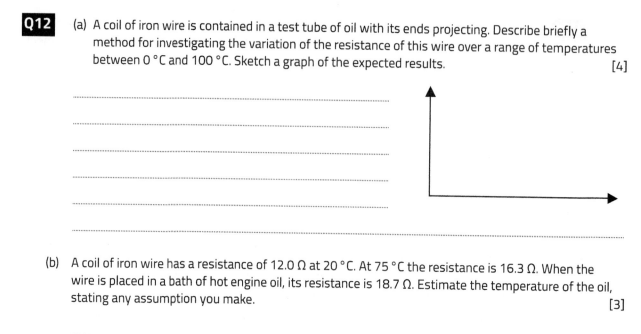

(b) A coil of iron wire has a resistance of 12.0 Ω at 20 °C. At 75 °C the resistance is 16.3 Ω. When the wire is placed in a bath of hot engine oil, its resistance is 18.7 Ω. Estimate the temperature of the oil, stating any assumption you make. [3]

Q13 *Point contact diodes* were developed in the 1940s for use in microwave receivers for radar. A website gives the following sketch graph of the variation of resistance with applied pd for a point contact diode.

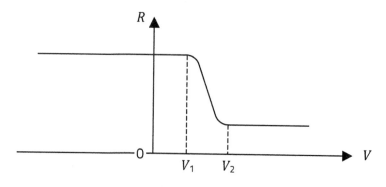

Sketch a current–voltage characteristic for a point contact diode, showing the features relating to V_1 and V_2. [3]

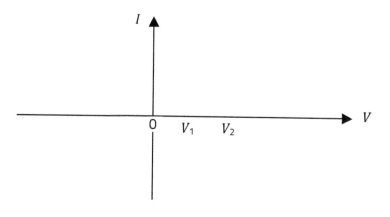

Question and mock answer analysis

Q&A 1 A group of students investigates the current–pd (*I–V*) graph of an old 12 V filament car headlamp. They obtain the following results.

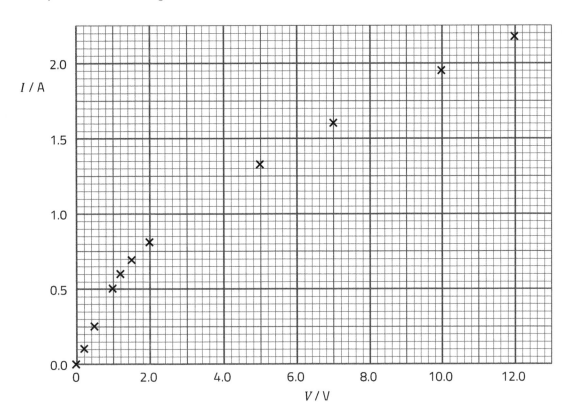

(a) With the aid of a circuit diagram, describe how they could have obtained these results. [3]

(b) From their results, describe how the **resistance** of the filament changes with pd. Include calculations in your answer. [4]

(c) Explain qualitatively, in terms of electrons, the variation of resistance with pd between 2 V and 12 V. [4]

(d) Use the graph to determine the pd at which this light bulb filament would dissipate the same power as a 4.0 Ω resistor and state the value of this power. [3]

What is being asked

Although the question is based upon a specified practical, only part (a) directly examines this, with a standard AO1 recall description. Part (b) looks like a standard piece of graph description. However there are two points to note, which make this an AO3 question: first, the variation of <u>resistance</u> is asked for, not current (i.e. the variable on the *y* axis); second, the graph has a straight line part and a curve. Part (c) is a standard piece of bookwork, albeit not easy, hence AO1. The comparison of the lamp with an ohmic resistor, (d), is a fairly straightforward part of evaluation (AO3).

Mark scheme

Question part			Description	AOs			Total	Skills	
				1	2	3		M	P
(a)			Circuit drawn with lamp connected across a power supply [1] Method of adjusting pd / current, e.g. rheostat or variable voltage supply [1] Ammeter in series and voltmeter in parallel with the lamp [1]	3			3		3
(b)			Between 0 and 1.2 V (or 0 and 0.6 A) the resistance is constant [1] at 2.0 Ω [1] At higher voltages (or currents) the resistance increases [steadily] [1] with identified value at specified voltage (or current), e.g. 5.5 Ω at 12 V [1] NB 3 max If no explicit calculation, e.g. $\frac{12\,V}{2.17\,A} = 5.5\,Ω$			4	4	2	
(c)			Resistance is caused by collisions between free (or conduction) electrons and metal atoms / ions / lattice [1] At higher currents, more energy is passed on in the collisions, raising the temperature [1] At higher temperature the time between collisions is less [1] so the drift velocity is lower and the current is therefore lower [than if the temperature were constant]. [1]	4			4		
(d)			Graph line drawn for (appropriate section of) the results and I–V graph for a 4 Ω resistor drawn (passing through 4.0, 1.0) [1] or equiv. Intersection of graphs identified – 5.8 V [1] Power = 8.4 W ecf [1]			3	3	1	
Total				7	0	7	14	3	3

Rhodri's answers

(a)

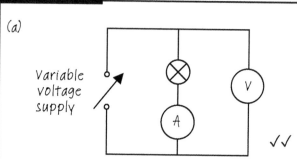

- Set up circuit
- Adjust voltage to zero – measure current
- Increase voltage in stages and read current each time. ✓

MARKER NOTE

A good answer which hits all the marking points

3 marks

(b) The resistance increases as the voltage increases. ✓

e.g. at 2 V,

$R = \frac{2}{0.8} = 2.5\ Ω$

At 12 V $R = \frac{12}{2.15} = 5.6\ Ω$ ✓

MARKER NOTE

Rhodri has missed the fact that the low voltage points lie on a straight line, giving a constant resistance. Hence he cannot gain the first two marks. He gains the other points because he correctly states the variation of resistance and calculates the value at 12 V.

2 marks

(c) At higher voltages and currents, the electrons are moving faster so the wire is at a higher temperature. ✓ The resistance of a metal wire increases with temperature, so the higher the voltage the greater the resistance.

(d)

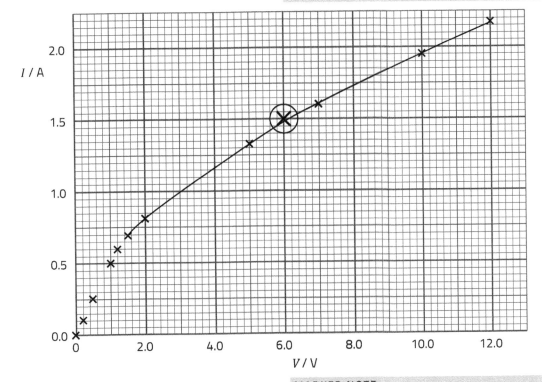

See graph for lamp

The bulb needs to have a resistance of 4 Ω – so 6 V, 1.5 A does this. ✗ [not enough]

$\text{Power} = \dfrac{V^2}{R} = \dfrac{6^2}{4} = 9 \text{ W}$ ✓ ecf

Ffion's answers

(a)

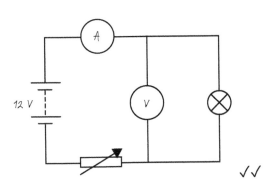

✓✓

In this circuit, set the variable resistor to its maximum resistance and note the current and voltage readings.

Adjust the variable resistor to a series of lower values and note the voltage and current readings. ✓

(b) Up to about 1.0 V, the graph is a straight line so the resistance is constant. ✓ Above 1.0 V, the graph curves to become more nearly horizontal so the resistance goes up with voltage. ✓

At 12 V the resistance is 5.5 Ω ✓ ✗

> **MARKER NOTE**
> A near perfect answer from Ffion. To obtain the 4th mark she needed to show the resistance calculation explicitly.
>
> **3 marks**

(c) At higher voltages, the electrons gain more kinetic energy between collisions with metal ions, so they pass on more energy, increasing the temperature. ✓ At higher temperatures because the random speed of the electrons is greater the time between collisions is less. ✓ Hence the mean drift speed is less and the current is less (than if the temperature was lower) ✓ so the resistance is higher. ✓

> **MARKER NOTE**
> Although this is essentially bookwork, the concepts are rather difficult and Ffion does well to put them across concisely.
>
> **4 marks**

(d)

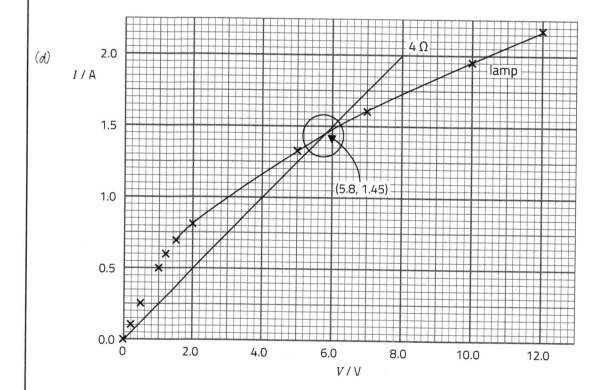

These are the I–V graphs for the lamp and a 4 Ω resistor. ✓

The two dissipate the same power where V and I are the same, i.e. when they cross (because P = VI). So the voltage is 5.8 V. ✓

So power = VI = 5.8 V × 1.45 A = 8.4 W ✓

> **MARKER NOTE**
> A well-explained answer from Ffion. She has drawn both graphs and correctly shown that the point of intersection has the relevant values of current and pd. She goes on to use one of the 3 possible methods of calculating power.
>
> **3 marks**

| Total | 13 marks /14 |

Section 3: DC Circuits

Topic summary

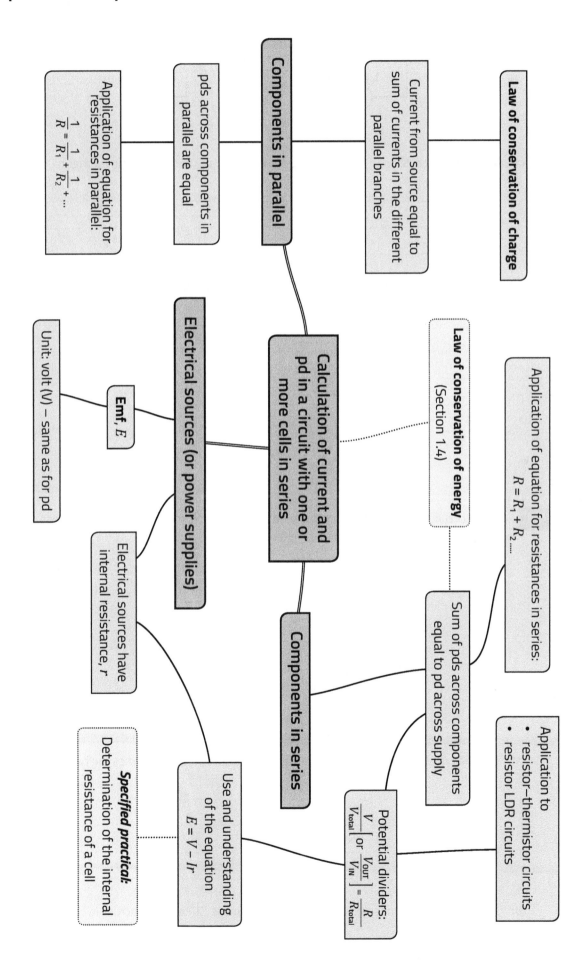

Components in parallel

Current from source equal to sum of currents in the different parallel branches

Law of conservation of charge

pds across components in parallel are equal

Application of equation for resistances in parallel:
$$\frac{1}{R} = \frac{1}{R_1} + \frac{1}{R_2} + \dots$$

Calculation of current and pd in a circuit with one or more cells in series

Electrical sources (or power supplies)

Emf, E

Unit: volt (V) – same as for pd

Electrical sources have internal resistance, r

Specified practical: Determination of the internal resistance of a cell

Use and understanding of the equation
$$E = V - Ir$$

Components in series

Law of conservation of energy (Section 1.4)

Sum of pds across components equal to pd across supply

Application of equation for resistances in series:
$$R = R_1 + R_2 \dots$$

- Application to
  - resistor–thermistor circuits
  - resistor LDR circuits

Potential dividers:
$$\frac{V}{V_{total}} \left[or \; \frac{V_{out}}{V_{IN}} \right] = \frac{R}{R_{total}}$$

Q1 In the circuit, the lamps $L_1 - L_3$ are identical. The greater the current, the brighter is the lamp.

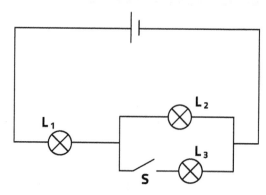

(a) With the switch, **S**, open as shown, explain why lamps L_1 and L_2 are equally bright. [2]

....

....

....

(b) Explain the changes in brightness of the three lamps when **S** is closed. [4]

....

....

....

....

....

....

Q2 In the circuit, the values of pd shown were obtained by connecting a voltmeter across each resistor. The battery has negligible internal resistance.

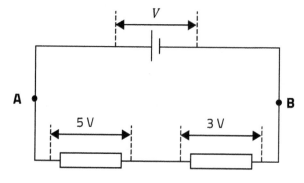

(a) Use the principle of conservation of energy to explain why the pd across the battery is 8 V. [2]

....

....

....

(b) A third resistor is connected directly between **A** and **B**. State the pd across it. .... [1]

Q3 You are provided with three 12 Ω resistors. What different values of current can you take from a 6.0 V power supply using different combinations of some or all of these resistors? You should sketch the combinations in the space below. [4]

Q4 A circuit is connected as shown. The power supply has negligible internal resistance.

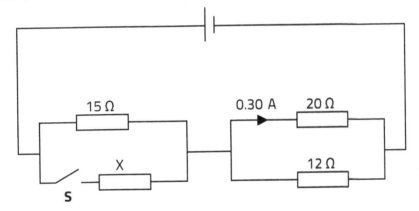

(a) Calculate the power dissipated in the circuit when the switch, **S**, is open as shown. [4]

....

....

....

....

....

....

(b) Without calculation, explain how the pd across the 15 Ω resistor changes when switch **S** is closed. [3]

....

....

....

....

....

....

Q5   The circuit shows the sensing circuit of a light alarm. The terminals labelled V_{OUT} are connected to the alarm input, which has a very high resistance.

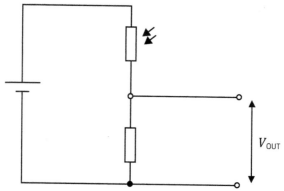

(a)   Explain how the output voltage, V_{OUT}, varies with light level incident upon the LDR.   [4]

....

....

....

....

....

....

(b)   The resistance–temperature graph for a thermistor is shown.

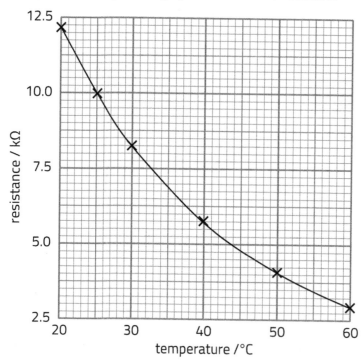

In the space to the right of the graph, **draw a sensing circuit** for a low-temperature alarm which should operate when the temperature drops below 37 °C. The power supply has a terminal pd of 12.0 V and the alarm is triggered when V_{OUT} from the sensor rises above 5.0 V.   [4]

Q6 The emf of an electric battery is 9.0 V.

(a) Explain what is meant by an emf of 9.0 V. [2]

(b) The battery produces a current of 1.5 A when it is connected into a circuit. The pd across the battery terminals is then 7.8 V.

(i) Account for the energy transfers in the battery and external circuit. [3]

(ii) Determine the internal resistance of the battery. [1]

Q7 A power supply has a terminal pd of 6.5 V when a 10 Ω resistor is connected across it. When a second 10 Ω resistor is connected in parallel with the first, the pd falls to 6.0 V. Determine the emf and internal resistance of the power supply. [4]

Q8 It can be shown that an electrical power supply transfers the maximum power to a circuit when the external resistance, R, is equal to the internal resistance.

A conventional (non-rechargeable) cell has an emf of 1.5 V and an internal resistance of 0.3 Ω. A rechargeable Ni-Cd cell has an emf of 1.2 V and an internal resistance of 35 mΩ. Compare the maximum power available from these cells. [4]

Q9 In a practical lesson, a student is given a sealed power supply consisting of a battery of unknown emf and resistor of unknown value, r, in series. They investigate the power supply using this circuit.

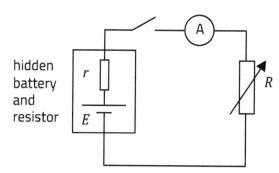

The variable resistance, R, is made by using some or all of the following resistors:

3.3 Ω, 10 Ω, 18 Ω

They obtained the following results:

R / Ω	3.3	6.4	7.6	10.0	13.3	18.0
I / A	0.43	0.34	0.31	0.27	0.23	0.19
$(I$ / A$)^{-1}$						

(a) Show how they used the available resistors to obtain the 7.6 Ω resistance. [2]

....

....

....

(b) Starting from the equations:

$$V = E - Ir \qquad \text{and} \qquad V = IR,$$

where V is the pd across the terminals of the power supply, show that a graph of $1/I$ against R should be a straight line of gradient $1 / E$. [3]

....

....

....

(c) Complete the table by adding in a row of values of I^{-1}, to an appropriate number of significant figures. [2]

(d) Use the grid on the next page to plot a graph of I^{-1} (on the y axis) against R. [4]

(e) The teacher tells the students that the emf of the battery is 4.8 V and the internal resistor has a value of 8.2 Ω. Evaluate whether this is consistent with the students' results. [4]

....

....

....

....

....

....

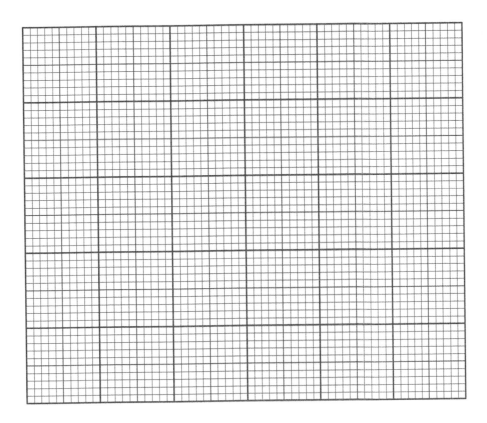

Q10 A 2.5 V, 1.5 W filament lamp is powered using 3 Ni-Cd cells, each of emf 1.2 V and negligible internal resistance, connected in series together with a resistor as shown in the circuit.

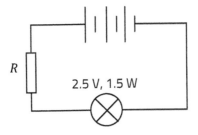

R

2.5 V, 1.5 W

(a) Calculate the resistance, R, required for the lamp to be powered at its rated value.   [3]

....

....

....

....

(b) Determine the fraction of the output power of the battery which is transferred within the resistor.   [2]

....

....

....

Q11 The silicon diode is a semiconductor device which, in normal operation, conducts in one direction only – shown in the following diagram of its symbol.

An approximate *I–V* characteristic for a diode is shown. Notice that, at the 'turn-on voltage' of 0.7 V, any value of current is possible.

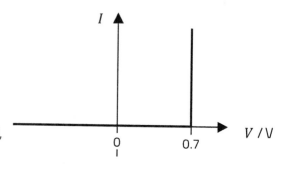

(a) A silicon diode is connected in series with a resistor of value 820 Ω. The two carry a current of 30 mA. Calculate the pd across the ends of the two. A diagram might help your answer. [2]

....

....

....

(b) A red light-emitting diode (LED) has a similar characteristic to the silicon diode. Its turn-on voltage is 1.9 V. The LED is to be used as an On-indicator for an electrical machine. The indicator circuit has a voltage supply of 9.0 V. The current in the LED must be between 10 mA and 25 mA in operation. Determine which of the following resistors are suitable for use as the series resistor. Give your reasoning.

68 Ω   100 Ω   220 Ω   470 Ω   680 Ω   1.0 kΩ   1.5 kΩ [3]

....

....

....

....

....

Question and mock answer analysis

Q&A 1 A battery has an emf of 9.0 V and internal resistance 1.8 Ω.

(a) Explain what is meant by 'an emf of 9.0 V'. [2]

(b) The battery is connected to a wire of resistance 5.4 Ω. Determine:

(i) The rate at which chemical energy is being transferred in the cell. [2]

(ii) The fraction of this energy which is transferred to the wire. [2]

(c) Sketch a graph of the terminal pd of this battery against the current delivered. Label the graph to show how particular features relate to the values of emf and internal resistance. [4]

What is being asked

This is quite a short question and seeks to determine whether you understand the concepts of emf and internal resistance. In part (a), the examiner could have asked, 'Explain what is meant by emf.' This would have been an AO1 question and required you to write the textbook definition. Asking it in this way requires you to use the data in an appropriate way and is therefore AO2. Part (b) has two calculations. which build on the answer to part (a). Again, it is AO2 because you need to apply your knowledge to a particular situation. Part (c) is a straightforward piece of bookwork, again applied to this battery. There are quite a few marks available, so it is clear that the examiner expects rather a lot of detail.

Unit 2 Practice questions

Mark scheme

Question part			Description	AOs			Total	Skills	
				1	**2**	**3**		**M**	**P**
(a)			9.0 J of energy are transferred from chemical to electrical [1] … per coulomb [of charge] {entering / passing through/ leaving} the battery [1]		2		2		
(b)	(i)		Current $\left[=\dfrac{9.0\,V}{5.4\,\Omega + 1.8\,\Omega}\right] = 1.25$ A [1] Rate of transfer [= 9.0 V × 1.25 A] = 11.25 W [1]		2		2	2	
	(ii)		Power dissipated in wire = $I^2 R = 8.43$ W [1] ecf Fraction $\left[=\dfrac{8.43\,W}{11.25\,W}\right] = 0.75$ [1] *Alternative answer* [Using potential divider] Fraction of total $V = \left[=\dfrac{5.4\,\Omega}{5.4\,\Omega + 1.8\,\Omega}\right] = 0.75$ (✓) I same in R and r so fraction of power = 0.75 (✓)		2		2	2	
(c)			Axes drawn and labelled, (V/V and I/A), straight-line graph sloping down from a +V intercept [1] V intercept labelled 9[.0] (accept E) [1] I intercept labelled 5[.0] (accept E / r) [1] [Accept statement that the maximum current is 5 or E / r] Gradient labelled -1.8 (accept -r) [1] Maximum mark for no correct number = 2 Maximum mark with only 1 correct number = 3	2	2		4	2	
Total				**2**	**8**	**0**	**10**	**6**	

Rhodri's answers

(a) The emf is the chemical energy lost per coulomb delivered by the battery ✗ ✓

MARKER NOTE

Rhodri's answer is a reasonable stab at the textbook definition of emf and gains the qualitative mark. He misses the mark which ties it in to the energy figure given.

1 mark

(b)(i) The current $= emf/tot\ resistance$

$= 9.0 / 7.2$

$= 1.25\ A$ ✓

Power in resistor $= I^2R = 1.25^2 \times 5.4$

$= 8.4\ W$

Power in cell $= I^2r = 1.25^2 \times 1.8$

$= 2.8\ W$

So total power $= 11.2\ W$ ✓

MARKER NOTE

Rhodri has not found the easiest way of answering this question but still gains both marks. He appears to realise that rate of energy transfer is power. After calculating the current, it would be much easier to use EI.

2 marks

(ii) The power in the wire $= \dfrac{V^2}{R} = \dfrac{9.0^2}{5.4}$ ✗

$= 15\ W$

So $\% = \dfrac{15}{11.2}\ \dfrac{11.2}{15} \times 100 = 74.6\%$ ✗

MARKER NOTE

Rhodri has appreciated that the wire is in series with the internal resistance. However, he imagines that the pd across the wire is the emf, so loses the first mark. His power in the wire is greater than the total power transfer, so he cannot rescue the calculation.

0 marks

(c)

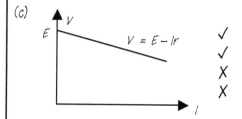

$V = E - Ir$ ✓ ✓ ✗ ✗

MARKER NOTE

Again, Rhodri hasn't spotted the need to use the data given and give the graph for *this* battery, so only achieves the first two marks.

2 marks

Total   **5 marks /10**

Ffion's answers

(a) This means that for every coulomb of charge which passes through the battery ✓, 9.0 J of chemical energy is used to push the charge around the circuit. ✓ [bod]

MARKER NOTE

Both marks are given. The examiner is not impressed by the 'energy used' but has given it the benefit of the doubt.

2 marks

(b)(i) $I = \dfrac{E}{R+r} = \dfrac{9.0}{5.4 + 1.8} = 1.25\ A$ ✓

Rate $=$ energy per second $= EI$

$= 9 \times 1.25 = 11.25\ W$ ✓

MARKER NOTE

An exemplary answer from Ffion, who realises that EI is the rate of energy transfer in the battery.

2 marks

(ii) pd across wire $= IR = 1.25 \times 5.4 = 6.75\ V$

So power in wire $= IV = 1.25 \times 6.75$

$= 8.4375\ W$ ✓

$\dfrac{8.4375}{11.25} \times 100\% = 75\%$ ✓

MARKER NOTE

Ffion has, unnecessarily, calculated the pd across the wire. She could have used this directly as follows:

$\dfrac{\text{power in wire}}{\text{total power}} = \dfrac{V_{wire}}{E} = \dfrac{6.75}{9.0} = 0.75$

But she still picks up both marks.

2 marks

(c)

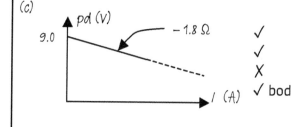

$- 1.8\ \Omega$ ✓ ✓ ✗ ✓ bod

MARKER NOTE

The only mark Ffion has not gained is the most difficult one – the intercept on the current axis. The 'bod' on the last mark is because the label on the graph doesn't clearly mention the gradient.

3 marks

Total   **9 marks /10**

Q&A 2 Sue and Ianto connect a resistance wire, **AB**, of length 75 cm and diameter 0.50 mm, into a circuit with a cell of emf 6.0 V and negligible internal resistance, as shown. They connect a voltmeter between **A** and a moveable point **P**.

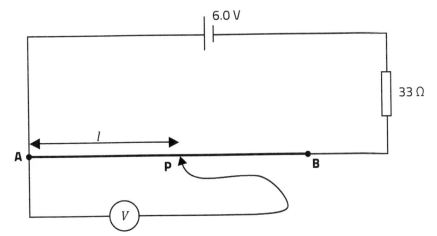

(a) Sue says that the voltmeter reading, V, should be proportional to the length, l, between **A** and **P**. Explain why she is correct. [Assume the voltmeter has infinite resistance.]   [3]

(b) Ianto obtains a set of results of V and l and plots a graph, which has a gradient of 0.020 V cm^{-1}. Use this value to determine the resistivity of the metal of the wire.   [5]

What is being asked

Part (a) of the question is a little unusual but it appears straightforward. You need to relate the variation of AP to potential divider ideas. It is not a standard explanation, hence it is AO3. In part (b) you can use the result of part (a), which is why the examiner asked you to show that Sue was correct, rather than whether she was correct. It then becomes a standard, if rather complex, AO2 calculation.

Mark scheme

Question part			Description	AOs			Total	Skills	
				1	2	3		M	P
(a)			There is no current through the voltmeter so the current is the same throughout AB [1] The resistance of AP is proportional to the length [1] Use of $V = IR$ or potential divider argument to show $V \propto l$ [1]			3	3		
(b)			$V_{AB} = 1.50$ V [1] pd across 33 Ω resistor = 4.5 V [1] Resistance of AB = 11.0 Ω [1] no ecf Rearrangement of $R = \dfrac{\rho l}{A}$ with r the subject at any stage [1] $\rho = 2.9 \times 10^{-6}$ Ω m ((**unit**)) [1] ecf on R_{AB}		5		5	4	
Total				0	5	3	8	4	

Rhodri's answers

(a) $R = \frac{\rho l}{A}$, so the resistance of the wire AP is proportional to the length ✓

So the voltage is also proportional to the length. ✗ not enough

MARKER NOTE

Rhodri has understood the importance of the current being the same at all points in the wire, so he can be awarded the first mark. To gain the last mark he needs clearly to tie in the voltage and resistance.

1 mark

(b) Current $= \frac{6.0}{33} = 0.182$ A ✗

$V_{AB} = 75 \times 0.020 = 1.5$ V ✓

Resistance of wire AB $= \frac{1.5}{0.182}$

$= 8.25 \, \Omega$ ✗

$\rho = \frac{RA}{l}$ ✓ ecf$= \frac{8.25 \times \pi (0.25 \times 10^{-3})^2}{0.75}$

So $\rho = 2.2 \times 10^{-6} \, \Omega \, m^{-1}$ ✗ unit

MARKER NOTE

Rhodri's first statement is incorrect. The 6 V pd is across the series combination of the wire and resistor. The pd across the 33 Ω resistor is 6.0 – 1.5 = 4.5 V. There is no ecf available for the resistance calculation. He rearranges the resistivity equation and obtains the mark but loses the last mark because of the incorrect unit.

2 marks

Total   **3 marks /8**

Ffion's answers

(a) The current is constant so the pd across the wire is constant ✗ [not clear]. If AP is half AB then the resistance of AP is half the resistance of AB ✓. The wire acts as a potential divider, so the pd across AP would be half that across AB. ✓

Similarly if AP is $\frac{1}{4}$ AB, the pd will be $\frac{1}{4}$ that across AB, so the pd is proportional to the length of AB.

MARKER NOTE

This is almost a perfect answer but the examiner doesn't feel able to award the first mark. What does 'constant' mean in this answer? Does it mean constant in time or constant in position? However, the other two marks are awarded.

2 marks

(b) pd across 1 cm of wire $= 0.02$ V

∴ pd across AB $= 75 \times 0.020 = 1.50$ V ✓

∴ pd across 33 W $= 6.0 - 1.5 = 4.5$ V ✓

∴ Current $= \frac{4.5}{33} = 0.136$ A

∴ $R_{AB} = \frac{1.5 \, V}{0.136 \, A} = 11 \, \Omega$ ✓

$R = \frac{\rho l}{A}$, $\rho = \frac{11 \times \pi (0.0005)^2}{0.75}$ ✓ [rearrange]

∴ $\rho = 1.2 \times 10^{-5} \, \Omega \, m$ ✗

MARKER NOTE

This is a well set-out answer. Ffion calculates the resistance of AB using the calculated current. She could also have used a potential divider argument. They are equally valid.

Her only mistake is to use the diameter of the wire instead of the radius, so she obtains the rearrangement mark but not the mark for the final answer.

4 marks

Total   **6 marks /8**

Section 4: The nature of waves

Topic summary

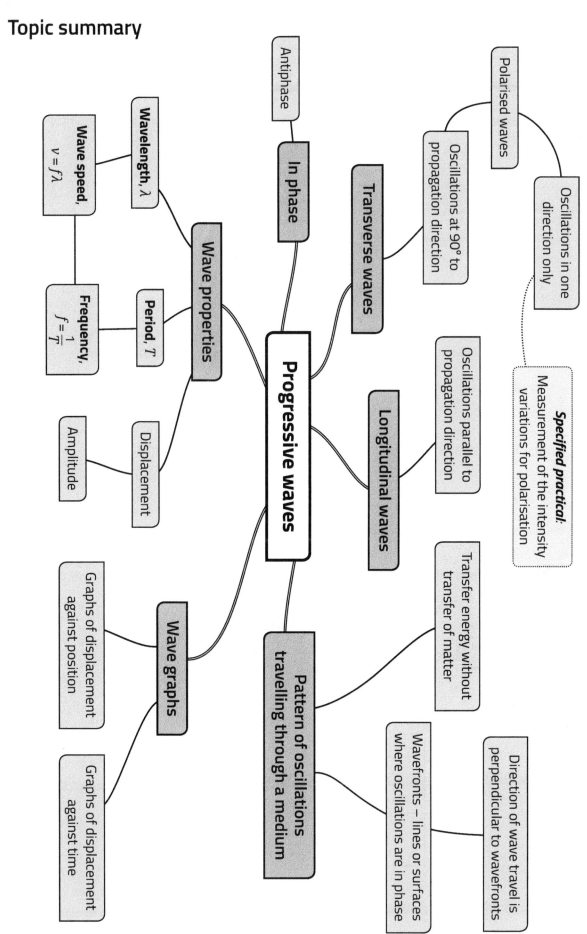

Q1 Energy can be transferred by moving fuel from one place to another. Explain briefly how energy transfer using progressive waves is different. [2]

....

....

....

Q2 Explain the difference between transverse and longitudinal waves and give an example of each. [4]

....

....

....

....

....

....

....

Q3 (a) Explain what is meant when light is described as being *polarised in one direction*. [1]

....

....

(b) Explain what is meant when light is described as being *partially polarised*. [2]

....

....

....

Q4 Unpolarised light of intensity 1.00 W m^{-2} is incident normally upon a slowly rotating polaroid and then a detector.

Detector   Unpolarised light of
                            intensity 1.00 W m^{-2}

                    Rotating
                    polaroid

(a) Explain why the graph shown of intensity against rotation angle is to be expected. [2]

intensity
/ W m^{-2}

1.0

0.5

                                        polaroid angle / °

....

....

....

(b)   Cheryl rotates a polaroid in a beam of light and obtains the following graph of intensity against polaroid angle.

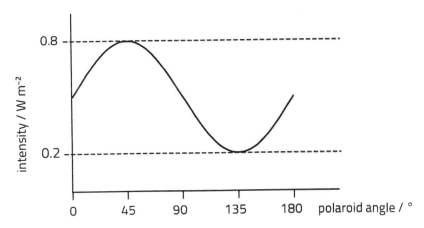

She suggests that the incident light beam has a total intensity of 1.0 W m^{-2}, with 60% polarised in one direction and the remainder unpolarised. Evaluate her suggestion.   [4]

....

....

....

....

....

....

....

Q5   Explain how an unpolarised light source and two polaroids can be used to investigate the polarisation of light and describe the expected observations.   [6 QER]

....

....

....

....

....

....

....

....

....

....

....

....

....

....

Q6 A sinusoidal wave travels in one direction on a long string.

3.04 m

At time $t = 0$, point A has a maximum displacement.

(a) Indicate a point on the string oscillating:
(i) in phase with A (label it B);
(ii) in antiphase with A (label it C). [2]

(b) A graph of the displacement of point A with respect to time is shown.

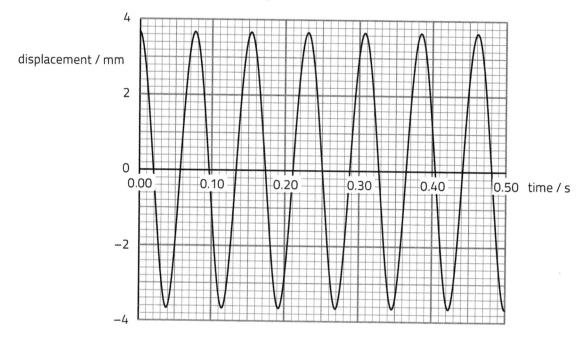

Calculate

(i) the wavelength, [1]

....

....

(ii) the amplitude, [1]

....

....

(iii) the speed of the wave. [3]

....

....

....

....

....

Q7 The diagram shows circular wavefronts that are peaks on the surface of water.

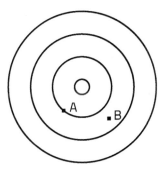

(a) Indicate on the diagram:

(i) the direction of propagation for point A; [1]

(ii) a distance corresponding to a wavelength (label it λ); [1]

(iii) a point oscillating in phase with point B (label it C); [1]

(iv) a point oscillating in antiphase with point B (label it D). [1]

(b) Joanna notices that 25 wavefronts pass point B in 8.0 s and that the distance between the first and fourth wavefronts is 18.0 cm. Calculate the speed of the wave. [3]

....

....

....

....

....

Q8 An earthquake strikes Wales with its epicentre at the Liberty Stadium in Swansea. The earthquake produces both longitudinal (P) and transverse waves (S). The P waves from the earthquake travel at a speed of 6.2 km s^{-1} and the S waves travel at a speed of 3.7 km s^{-1}. Both the S and P waves have a frequency of 8.9 Hz.

(a) Calculate the wavelengths of the P and S waves. [3]

....

....

....

....

....

(b) In Bala, a North Wales town, a seismogram reveals that there is a delay of 16.3 s between the arrival of the longitudinal waves and the transverse waves. Calculate the distance between Swansea and Bala. [3]

....

....

....

....

....

Q9 Waves travel along the surface of water in a ripple tank. The wavefronts and direction of propagation of the waves can be seen in the diagram.

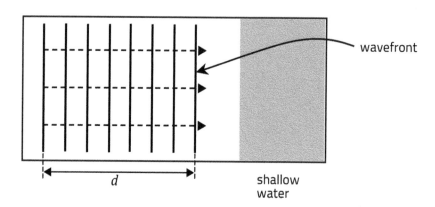

(a) State the relationship between the direction of propagation, the line of wavefronts and the direction of oscillation of the water surface. [2]

....

....

....

(b) Gerallt measures the distance, d, shown in the diagram as 14.6 cm and also counts 20 waves passing a certain point in 4.7 s. Calculate the speed of the waves. [4]

....

....

....

....

....

....

....

(c) The waves now pass to the shallow water where the propagation speed is less. Gerallt states that the frequency of the waves stays the same and their wavelengths shorten. Evaluate Gerallt's conclusions. [3]

....

....

....

....

Question and mock answer analysis

Q&A 1

(a) Explain the difference between the terms *displacement* and *amplitude* for a wave. [2]

(b) State a definition of the term *wavelength* that applies to both transverse and longitudinal waves. [2]

(c) The displacement of waves is obtained at varying distances from the source at time $t = 0$ using a photograph. A graph of displacement against distance is then produced.

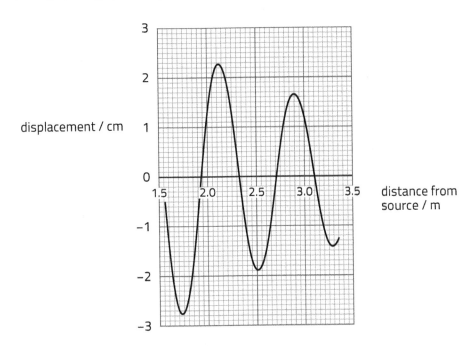

(i) Describe briefly the variation in amplitude and explain why this occurs. [2]

(ii) The frequency of oscillation of the wave is 44 Hz. Sketch a graph of the variation of the displacement of the wave with time at 2.9 m from the source on the grid below.

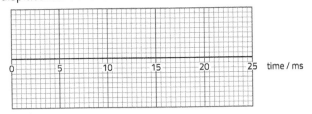

[4]

(iii) Calculate the speed of the waves. [3]

(iv) The intensity of the wave is proportional to the amplitude squared. Evaluate whether or not the variation in intensity follows the inverse square law $\left(\text{intensity} \propto \dfrac{1}{\text{distance}^2}\right)$ [4]

What is being asked

Parts (a) and (b) are almost standard definitions but have a slight twist. Nonetheless, they are still AO1 marks. Part (c)(i) requires a simple description of the results and an explanation for this description. Although evaluation of results is usually AO3, this is more of a quick piece of analysis and a standard explanation for the amplitude drop. Parts c(ii) & (iii) are also based on analysis and calculation and hence are AO2 marks. Part c(iv), on the other hand, involves quite tricky AO3 marks because checking for the inverse square law requires a good understanding and strategy.

Mark scheme

Question part		Description	AOs			Total	Skills	
			1	2	3		M	P
(a)		Displacement is the distance from the equilibrium position [1] Amplitude is the maximum displacement [1]	2			2		
(b)		The (minimum) distance between two adjacent points [1] Oscillating in phase (but not on the same wavefront) [1]	2			2		
(c)	(i)	Amplitude decreases with distance [1] Due to wave spreading out in 2D/3D OR accept due to resistive forces/damping [1]		2		2		
	(ii)	Period calculated or implied (22.7 ms) [1] Amplitude is 1.65 cm. Accept 1.6 − 1.7 (cm) [1] y-axis labelled (numbers, title, unit) [1] Reasonable freehand sinusoid; correct phase (cos wave), period and amplitude in plot [1]		4		4	3	
	(iii)	Obtaining wavelength = 0.75 ± 0.05 (m) [1] Using equation $v = f\lambda$ [1] Answer = 33 (m s^{-1}) [1]	1	1 1		3	2	
	(iv)	Selecting 1 set of appropriate data, e.g. dist = 1.75, A = 2.75, dist = 3.25, A = 1.45 etc. [1] Realising amplitude$^2 \propto \dfrac{1}{\text{distance}^2}$ [1] Choosing an appropriate method e.g. checking Ad = constant OR A^2d^2 = constant OR calculating a second amplitude/distance using $A = \text{constant}/d$ [1] Result and valid conclusion e.g. 4.76, 4.71 quite close (allow ecf) [1]			4	4	4	
Total			5	8	4	17	9	

Rhodri's answers

(a) Displacement is the instantaneous height of the wave (needs more, no bod) whereas amplitude is the biggest value that this can take. ✓

MARKER NOTE
Height of the wave is not good enough for the displacement especially if the wave is longitudinal. However, 'the biggest value it can take' is equivalent to the maximum in the mark scheme.
1 mark

(b) The wavelength is the distance between consecutive crests or compressions. ✓ bod

MARKER NOTE
Rhodri has been crafty and has come up with an answer that almost satisfies the demand of the question. He does not know this standard definition but his answer satisfies the condition for the first mark, i.e. 'the distance between consecutive crests or compressions' is equivalent to 'The distance between two adjacent points'.
1 mark

(c) (i) The amplitude is clearly decreasing over time and so this is damped oscillations. ✓ bod

MARKER NOTE
Rhodri has completely misunderstood what is going on. He thinks that the amplitude is decreasing over time when it is decreasing with distance from the source. Nonetheless, his point about damping hits the 2nd marking point and is awarded the 2nd mark with bod.
1 mark

(ii) $T = \dfrac{1}{f} = \dfrac{1}{44}$ ✓

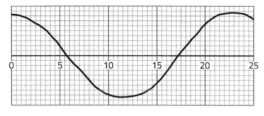

✓ bod for amplitude (no scale)

✓

MARKER NOTE

Here, Rhodri's final graph is better than Ffion's but he obtains fewer marks because he does not explain clearly what he is doing. The 1st mark is deserved for the period, especially when the graph is examined (leaving the answer as a fraction is not a good idea). The 2nd mark is awarded with a very generous bod because the amplitude is implied from the graph (even though no y-axis scales are shown). The 3rd mark cannot be awarded because the y-axis is unlabelled. The 4th mark is deserved for a good plot but still requires bod because the actual amplitude is implied rather than shown clearly to be correct.

3 marks

(iii) 36 m s^{-1} No bod

MARKER NOTE

Rhodri's risky tactic of not showing his workings has backfired here. He has gained zero marks even though his answer is possibly correct (see Ffion's answer). His speed of 36 m s^{-1} suggests a wavelength of 0.82 m which is outside tolerance and the examiner is left with no option but to award zero marks. Had Rhodri written 35 m s^{-1} he might well have received full marks but, as it appears, there is no evidence of any correct workings.

0 marks

(iv) Are you `avin´ a laugh? What does inverse square law mean? My teacher hasn´t taught me this! I reckon the statement is correct because it usually is. If you look at the numbers, when the amplitude goes down to 74% (2.3 to 1.7) the distance increases ✓ from 2.1 to 2.9 (no bod) which is sort of the same ratio (2.1 is 72% of 2.9). ✓ (bod)

MARKER NOTE

First, the examiner has to smile at Rhodri's silly comments and then ignore them (as long as they are not serious enough to be reported). Then, the examiner realises that Rhodri is very close to obtaining full marks even though it is possible that he doesn't know what he is doing. The 1st mark is obviously deserved but the 2nd mark is definitely not awardable. The 3rd mark is difficult to judge because Rhodri is, essentially, checking that the amplitude and distance are inversely proportional but the examiner has decided against awarding bod. The 4th mark is awarded because the correct conclusion is reached (just) and this method is valid even though Rhodri has not explained (or understood) this method. Note that this is an excellent example of perseverance leading to marks – many candidates would have left this part blank and received nothing.

2 marks

Total	8 marks /17

Unit 2 Practice questions

Ffion's answers

(a) Displacement is a vector and is defined as the distance from the equilibrium position ✓ and amplitude is the maximum displacement. ✓

MARKER NOTE

Ffion's answer is even better than the mark scheme because she has added that the displacement is a vector.

2 marks

(b) The wavelength is the distance between the two nearest points ✓ bod oscillating in phase. ✓

MARKER NOTE

Ffion's answer has received the benefit of doubt for the 1st mark but any examiner who understands English would have awarded the mark. Ffion has replaced the word 'adjacent' with 'nearest' which should be acceptable to any examiner. Note that the word 'minimum' is in brackets in the mark scheme. This means that the word 'minimum' is an optional extra and is not compulsory.

2 marks

(c) (i) The amplitude is decreasing as you get further and further away from the source ✓ – this is the inverse square law. No bod

(ii) To calculate the period $T = \frac{1}{f} = 0.023\ s$ ✓

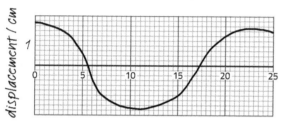

The amplitude is 1.65 cm ✓

✓ bod amplitude looks slightly out at the top

✓ bod only one number on axis

(iii) wavelength = 0.8 m ✓

Speed = 44 × 0.8 = 35.2 m/s ✓✓

(iv) The amplitude at distance 1.75 is 2.85 cm while the amplitude at 3.275 is 1.45 cm. ✓

If $I \propto 1/d^2$ then $I = k / d^2$

Using first data $k = I\,d^2 = 1.75 \times 2.85^2 = 14.2$

Using 2nd data $k = I\,d^2 = 3.275 \times 1.45^2 = 6.9$

Hence, I conclude that the intensity does not follow the inverse square law. ✓ ecf

Total **14 marks /17**

Section 5: Wave properties

Topic summary

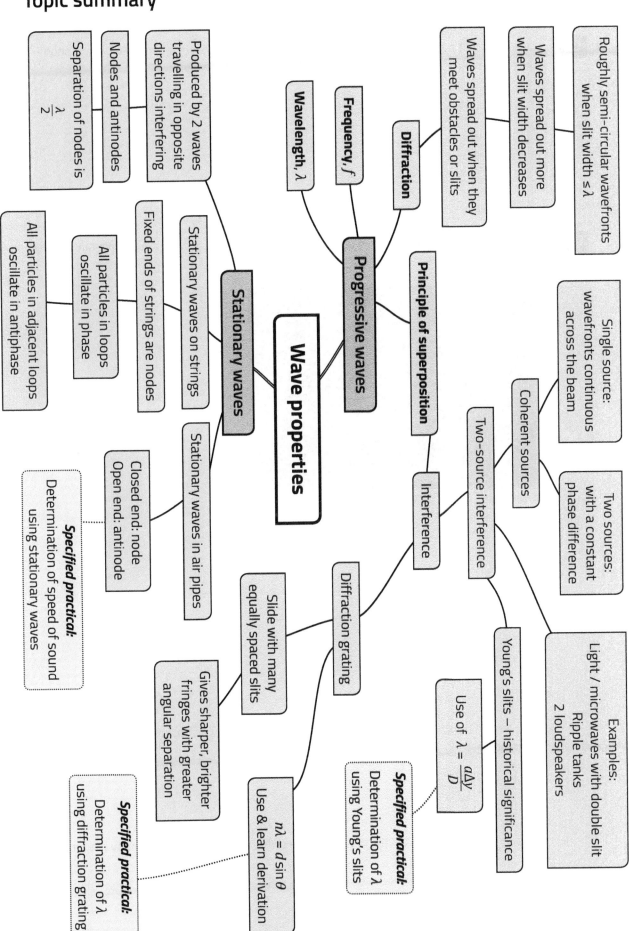

- **Wave properties**

- **Progressive waves**
  - **Wavelength, λ**
  - **Frequency, f**
  - **Diffraction**
    - Waves spread out when they meet obstacles or slits
    - Waves spread out more when slit width decreases
    - Roughly semi-circular wavefronts when slit width $\leq \lambda$
  - **Principle of superposition**
    - Interference
      - Two-source interference
        - Coherent sources
          - Single source: wavefronts continuous across the beam
          - Two sources: with a constant phase difference
        - Examples: Light / microwaves with double slit, Ripple tanks, 2 loudspeakers
        - Young's slits – historical significance
        - Use of $\lambda = \dfrac{a\Delta y}{D}$
        - *Specified practical:* Determination of λ using Young's slits
      - Diffraction grating
        - Slide with many equally spaced slits
        - Gives sharper, brighter fringes with greater angular separation
        - $n\lambda = d\sin\theta$ Use & learn derivation
        - *Specified practical:* Determination of λ using diffraction grating

- **Stationary waves**
  - Produced by 2 waves travelling in opposite directions interfering
  - Nodes and antinodes
  - Separation of nodes is $\dfrac{\lambda}{2}$
  - Stationary waves on strings
    - Fixed ends of strings are nodes
    - All particles in loops oscillate in phase
    - All particles in adjacent loops oscillate in antiphase
  - Stationary waves in air pipes
    - Closed end: node Open end: antinode
    - *Specified practical:* Determination of speed of sound using stationary waves

Q1 *Diffraction* is a phenomenon that occurs in waves. Explain briefly what is meant by *diffraction*. [2]

....

....

Q2 Describe what happens to the diffraction pattern when light of wavelength 600 nm is shone upon a single slit whose width is increased gradually from 300 nm to 6 µm. [3]

....

....

....

....

....

Q3 (a) The image shows a two-source interference pattern for water waves in a ripple tank. Explain how this pattern occurs. [3]

....

....

....

....

....

....

(b) The path difference between waves from the two sources is $\frac{1}{2}\lambda$ at point A. State what the path difference is:

(i) for point B .... [1]

(ii) for point C .... [1]

(c) The two sources of waves in the ripple tank are *coherent*. State what is meant by the term *coherent*. [1]

....

....

Q4 (a) State the *principle of superposition*. [2]

....

....

....

(b) (i) Explain what is meant by the terms *constructive* and *destructive interference*. [3]

....

....

....

....

(ii) Two waves of equal amplitude but different frequencies meet. Their time-varying displacements are shown in the graph. Use the principle of superposition to obtain the net displacement at times 0 s, 0.35 s, 0.5 s, 0.65 s and 1.0 s. Hence sketch the resulting waveform on the graph. [4]

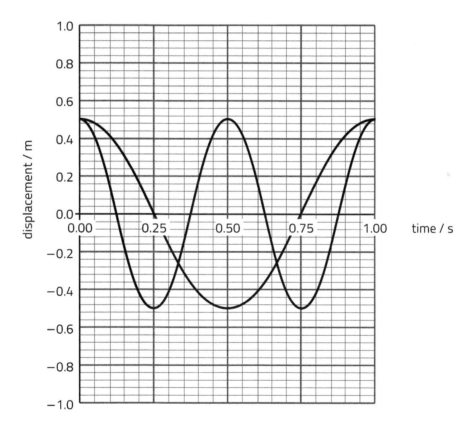

Q5 (a) State the historical significance of Young's double slit experiment. [1]

....

....

(b) Young's double slit experiment is an experiment that can be carried out in a school laboratory using a laser and double slit. Explain how the experiment should be carried out and how a suitable graph will lead to a value of the wavelength of the laser light. [6 QER]

....

....

....

....

....

....

....

....

....

....

....

....

....

....

....

....

....

....

....

....

....

....

Q6 Two-source interference is observed using microwaves. The microwave source is the same distance from the slits S_1 and S_2. So is the point P.

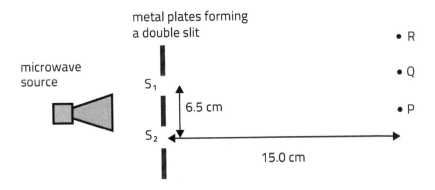

(a) Explain why a maximum signal is observed when a microwave detector is placed at point P. [3]

....

....

....

....

(b) Q is the first point above P where a minimum signal is detected and R is the first point above P where a maximum signal is detected.

State and explain the values of $S_2Q - S_1Q$ and $S_2R - S_1R$ in terms of the wavelength, λ, of the microwaves.. [2]

....

....

....

(c) (i) The wavelength of the microwaves is 2.8 cm.

Use the equation: $\lambda = \dfrac{a\Delta y}{D}$

to calculate the distance between point P and point Q. [2]

....

....

....

(ii) Evaluate whether or not the equation $\lambda = \dfrac{a\Delta y}{D}$ is a good approximation for the calculation in part (i). [2]

....

....

....

....

Q7 (a) Derive the equation for a diffraction grating. You may add to the diagram. [3]

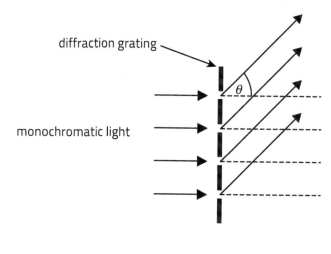

(b) In an experiment to measure the wavelength of laser light using a diffraction grating, Meinir obtains the following results (the $n = -1, 0, +1$ dots are shown):

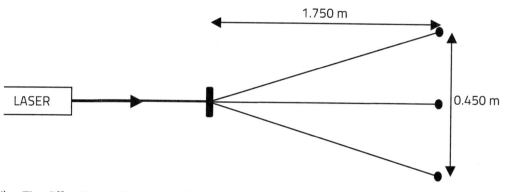

(i) The diffraction grating states that it has 250 lines/mm. Calculate a value for the wavelength of the laser light. [3]

(ii) Calculate the total number of bright dots produced by this diffraction grating when used with this laser. [3]

Q8 Meurig carries out an experiment to measure the speed of sound in air using the following apparatus.

(a) Show the shape of the first harmonic in the diagram. [1]

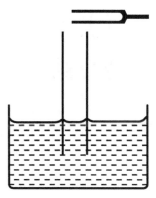

(b) Show that the speed of sound, c, is related to the length, l, of the 1st harmonic by the relationship:

$$c = 4lf$$

where f is the frequency of the tuning fork. [2]

....

....

....

(c) Complete Meurig's table of results and discuss the accuracy of his results given that the true value of the speed of sound is 342 ms^{-1} at the temperature of the laboratory. [4]

frequency / Hz	length / cm	Speed / m s^{-1}
256	31.2	319
288	27.5	
320	24.6	315
384	20.1	
427	17.9	306
480	15.7	301

....

....

....

....

(d) Rachel suggests that Meurig should have drawn a straight-line graph and used the gradient to obtain the speed of sound. Evaluate to what extent Rachel is correct. [4]

....

....

....

....

....

....

....

Question and mock answer analysis

Q&A 1

(a) Explain the difference between progressive and stationary waves in terms of energy, amplitude and phase. [4]

(b) (i) The metal walls inside a microwave oven reflect microwaves. Suggest why stationary waves are produced inside a microwave oven. [2]

    (ii) Brynley melts a bar of chocolate in a microwave oven with no rotating table and notices that the chocolate bar melts in positions that are separated by 6.1 cm. Calculate the frequency of the microwaves. [3]

(c) Blodeuwedd carries out an experiment to investigate the polarisation of a microwave source and adopts the following set-up.

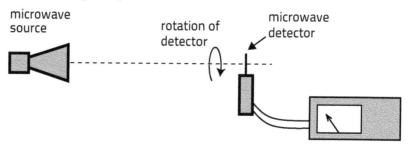

As the detector is rotated, the signal starts from a maximum and drops to zero when the detector has been rotated 90°.

Explain these observations. [4]

What is being asked

There are many, many questions that an examiner can ask within this rather large and popular topic (for examiners). This question is about stationary waves, progressive waves and microwaves. The last part of the question is slightly synoptic because it relates back to Section 2.4 – The nature of waves, and polarisation in particular. Part (a) is a standard question asking about the differences between progressive and stationary waves. Although slightly tricky, this is straight off the specification and the marks are then classified as AO1. Part (b)(i) is asking about how stationary waves are formed in a slightly unusual context and hence an analysis of what is occurring is required – these then are AO2 marks. Part (b)(ii) is a two-part calculation based on stationary waves. These are mainly AO2 marks but the simple use of the wave formula is only considered to be AO1. The final part (c) requires a slightly tricky explanation and is based on polarisation of microwaves. Here, candidates are required to make judgements and reach conclusions in a novel situation, which elevates these marks to AO3 skills.

Mark scheme

Question part			Description	AOs			Total	Skills	
				1	2	3		M	P
(a)			Progressive waves transfer energy, stationary waves do not [1]	4			4		
			Amplitude is constant (or decreases from source) for progressive but varies from max (antinode) to zero/min (node) in stationary [1]						
			Phase is constant within loop but antiphase for adjacent loops for stationary [1]						
			Phase gradually increases/changes with distance for progressive [1]						
(b)	(i)		Reflected waves interfere [1] Antinodes – constructive OR nodes – destructive [1]		2		2		

	(ii)	Distance between nodes/antinodes = $\lambda/2$ or implied [1] Use of $c = f\lambda$ e.g. $f = \dfrac{3 \times 10^8}{0.122}$ [1] Final answer = 2.46 GHz [1]	1	1	1 1	3	2
(c)		Microwave source is polarised [1] Source & detector initially aligned giving max [1] At 90° waves from source cannot be detected [1] Explanation of either: • why microwave source is polarised OR • why detector detects only one direction of polarisation [1] e.g. electric field/current goes back and forth in one direction only (in both source and detector)			4	4	
Total			**5**	**4**	**4**	**13**	**2**

Rhodri's answers

(a) Progressive waves transfer energy ✓ [bod], have a constant amplitude (but stationary varies) and the phase is continually changing.

MARKER NOTE

Rhodri's response to the energy aspect makes no mention of stationary waves not transferring energy – this is very risky. However, the examiner has awarded the mark on this occasion because the second part of the explanation is implied. Rhodri cannot gain the amplitude mark because he makes no mention of nodes and antinodes. The 3rd and 4th marks cannot be awarded either because 'the phase is continually changing' doesn't clearly mention the waves to which it refers and wouldn't be enough even if it mentioned progressive waves because there is no mention of distance from the source.

1 mark

(b) (i) As the microwaves bounce back and forth in the 'popty ping' they interfere ✓ with each other and cause nodes and antinodes.

MARKER NOTE

Rhodri's answer is quite good and he has provided a bit of entertainment for the examiner by using the Welsh slang for a microwave oven (popty-ping). He obtains the 1st mark but cannot obtain the 2nd mark because he has not mentioned constructive/destructive interference.

1 mark

(ii) Using $c = f\lambda$,

$$f = \frac{c}{\lambda} = \frac{3.0 \times 10^8}{6.1} \checkmark$$

$$= 49.2 \text{ MHz } ✗$$

MARKER NOTE

Rhodri cannot obtain the 1st mark because he believes the separation of nodes/antinodes to be a whole wavelength but he can obtain the 2nd mark for the use of the equation. There is no ecf applicable to obtain the last mark here. Notice that he has kept the distance in cm but this would have been penalised in the last mark, which he has already lost!

1 mark

(c) At the start the polaroids ✗ are aligned and you get a maximum signal. As you rotate the second polaroid ✗, the signal drops gradually to zero at 90° when you have cross–polaroids. ✓[ecf]

MARKER NOTE

Rhodri's answer is a reasonable attempt but he is talking about two polaroids when there are none in this question. He loses the 1st mark because he does not mention the source. He also loses the 2nd mark because he has mentioned neither the detector nor the source. The 3rd mark is awarded through ecf (as would have been agreed in the markers' conference) but the last mark is extremely difficult and cannot be awarded.

1 mark

Total   **4 marks /13**

Ffion's answers

(a) Stationary waves do not transfer energy whereas progressive do ✓ (without transferring matter). Wave amplitude drops off as inverse square for progressive waves but varies from max to zero between antinodes and nodes ✓. For a progressive wave, points lag more and more the further they are from the source ✓ but the phase is always constant inside a loop for a stationary wave.

MARKER NOTE

Ffion's energy explanation is fine and obtains the mark. Her statement regarding the amplitude of both types of waves also merits the mark. However, note that the amplitude of waves does not drop off as inverse square – it is the intensity that tends to drop off as inverse square (this is not penalised here because it is extra detail that is not required). Ffion's explanation of the phase of a progressive wave is a rare example of a correct answer but she stops short of providing enough information regarding the phase of a stationary wave – she omitted to mention that adjacent loops are in antiphase.

3 marks

(b) (i) Reflected waves will interfere ✓ with waves travelling in the opposite direction leading to areas of constructive interference (nodes) and destructive interference (antinodes). In the oven cavity there could be a 3D grid of nodes and antinodes because of the reflections off all 6 walls.

MARKER NOTE

Ffion's answer is exemplary but she has mixed up the nodes and antinodes and cannot obtain the 2nd mark. Her statement regarding the 3D pattern, although awesome, cannot be rewarded because it is not on the mark scheme.

1 mark

(ii) Wavelength = 6.1 × 2 = 12.2 cm ✓

$$f = \frac{c}{\lambda} = \frac{3.0 \times 10^8}{12.2} \checkmark$$

$$= 2.46 \times 10^7 \text{ Hz}$$

MARKER NOTE

Ffion clearly earns the first two marks but cannot obtain the final mark because she has forgotten to convert the wavelength from cm to m. This is a common slip but unusual for a very good candidate like Ffion.

2 marks

(c) First, the source must be emitting polarised waves for this effect to be observed ✓. Presumably, the detector only detects one direction of polarisation and no polaroid is needed. At the start, the source and detector are aligned and a max signal ✓ is obtained. This drops gradually to zero when the detector is rotated 90° because the detector then is at 90° to the polarisation of the source ✓, so the source has no component in the direction that the detector can detect.

MARKER NOTE

Ffion has made a superb attempt at this difficult question and clearly merits the first three marks. Her comment *'Presumably, the detector only detects one direction of polarisation and no polaroid is needed'* is insightful but falls short of the explanation required by this tough mark scheme. Her explanation that, at 90°, one polarisation has no component in the other direction is also superb but cannot be rewarded.

3 marks

Total	9 marks /13

Section 6: Refraction of light

Topic summary

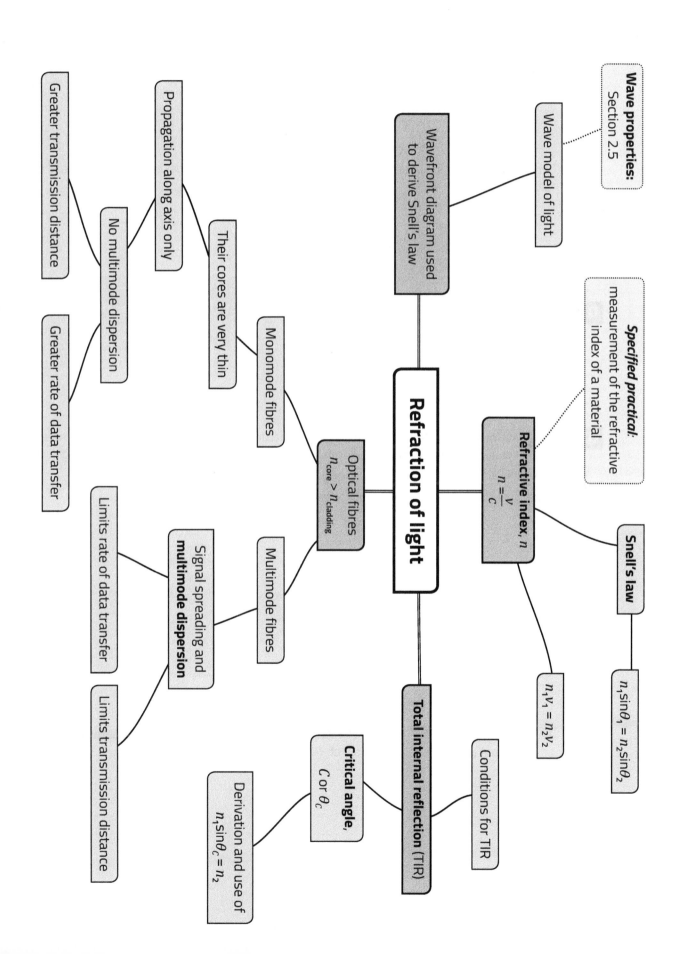

Q1 (a) State Snell's law. [2]

(b) Use Snell's law to define the term *refractive index* of a material in terms of angles. [2]

(c) Jennifer said that there was an equally good definition of *refractive index* in terms of the speed of light. Give this definition. [2]

Q2 Calculate the speed of light in a medium of refractive index 1.49. [2]

Q3 *Cherenkov radiation* is a form of radiation that can occur when electrons travel faster than the speed of light within a particular medium – it can be seen clearly in the cooling water surrounding nuclear reactors.

(a) State why Cherenkov radiation is impossible if the 'medium' is a vacuum. [1]

(b) (i) Calculate the minimum speed of electrons in water that will produce Cherenkov radiation ($n = 1.33$ for water). [2]

(ii) Calculate the pd required to accelerate an electron to this speed. Assume that the high velocity does not require Einstein's Theory of Relativity. [2]

Q4 A beam of light passes from air through a parallel-sided glass sheet ($n = 1.52$) and into the water ($n = 1.33$) of a fish tank, as shown. The initial angle of incidence is 45.0°. The angles are not drawn accurately.

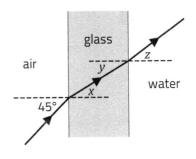

(a) Without calculation, explain why the light changes direction as shown at the two surfaces. [2]

....

....

....

(b) Calculate angles x, y and z. [3]

....

....

....

....

....

....

(c) Jordan says that you can calculate the angle z without first calculating y. Evaluate this statement without calculations. [2]

....

....

....

Q5 Light is incident upon a sphere of refractive index 1.42 as shown (the normal at the point of incidence has been added to the diagram).

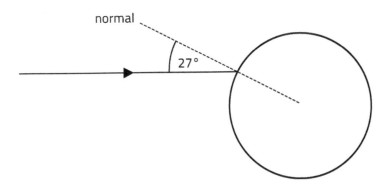

(a) Calculate the refracted angle as light enters the sphere **and draw it in the diagram**. [3]

....

....

....

(b) Draw two more rays to show the light being **reflected and refracted** at the point where the light ray you have drawn in part (a) is incident upon the opposite side of the sphere. [2]

Q6 Light in air is incident upon a glass prism of refractive index 1.55 at an angle, i, to the normal of the front surface. The prism's cross-section is an equilateral triangle (see below). The angles i and c are not drawn with their correct values.

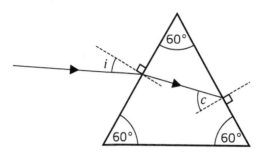

The light refracts at the front surface of the mirror but when it is incident upon the second surface, it is incident at the critical angle, c, of the glass.

(a) State the conditions required for total internal reflection. [2]

....

....

....

(b) Calculate the critical angle of the glass. [2]

....

....

....

(c) Calculate the incident angle, i. [3]

....

....

....

....

(d) Briony claims that, because the light is incident on the second surface at the critical angle, the light will never leave the prism and will be totally internally reflected each time the ray is incident upon one of the prism's surfaces. Determine whether, or not, Briony is correct. [3]

....

....

....

....

....

Q7 Explain how you would carry out an experiment to measure the refractive index of a glass block. You should also refer to your graphical method of analysing your results. [6 QER]

....

....

....

....

....

....

....

....

....

....

....

....

....

Q8 A student designs an alarm system to detect when the level of benzene ($n = 1.51$) in a tank drops below the minimum height. The arrangement is shown in the diagram. The prism is fixed in position and made of glass with a refractive index of 1.50.

Explain how the system is meant to operate. [4]

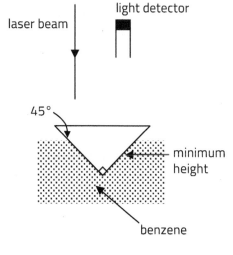

....

....

....

....

....

....

Q9 Light is incident upon an optical fibre as is shown in the diagram.

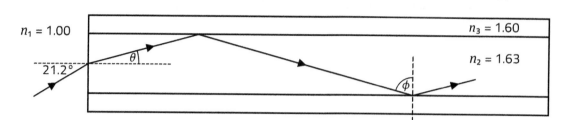

(a) Calculate the angles θ and ϕ. [3]

....

....

....

....

(b) Explain why light transmitted at this angle will not undergo total internal reflection. [3]

....

....

....

....

....

(c) The ray shown in the diagram below <u>does</u> undergo total internal reflection.

(i) Calculate the extra distance this light ray travels along a 14.0 km optical fibre compared with a light ray that travels straight along the axis of the fibre. [2]

....

....

....

(ii) Hence, explain the term *multi-mode dispersion*. [3]

....

....

....

....

Q10 An isosceles right-angled prism, **ABC**, is made from glass of refractive index 1.60. The face, **AC**, of the prism is horizontal. A narrow horizontal beam of light is incident from air on the face **AB** of the prism as shown in the diagram. The refracted beam is subsequently incident on face **AC**.

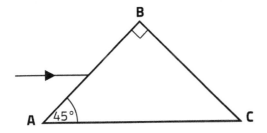

(a) Show that the light is totally reflected at face **AC** and subsequently emerges from face **BC** parallel to its original direction. Add to the diagram to illustrate this. [5]

....

....

....

....

....

....

(b) A second light beam, parallel to the first, is incident on face **AB** nearer to **A** than the first beam. Add the path of this light ray to the diagram and suggest why such a prism can be used as an inverting prism (i.e. objects viewed through it appear upside down). [3]

....

....

....

(c) James said that, whatever the refractive index of the glass, the angle of incidence on face **AC** will always be greater than the critical angle. Hence it doesn't matter what the refractive index of the glass is, in this inverting prism arrangement. Discuss to what extent he is correct. [4]

....

....

....

....

....

....

....

....

Question and mock answer analysis

Q&A 1 Geraint carries out an experiment to measure the refractive index of a glass block. He employs the following apparatus and records his results in the following table.

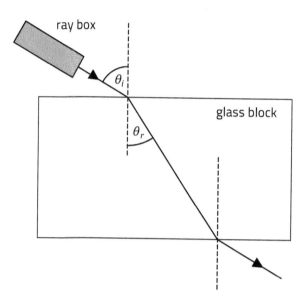

$\theta_i/°$	$\theta_r/°$	$\sin\theta_i$	$\sin\theta_r$
0	0	0	0
10	7	0.17	0.12
20	13	0.34	0.22
30	20	0.50	0.34
40	26	0.64	
50	31		0.52
60	36	0.87	0.59
70	39	0.94	0.63
80		0.98	0.66

(a) Complete the table. [3]

(b) Plot a graph of $\sin\theta_r$ (on the y axis) against $\sin\theta_i$ (on the x-axis), making best use of the grid, and draw a line of best fit. [Note: Page of graph grid provided, 12 × 9 squares] [5]

(c) Discuss to what extent the graph confirms Snell's law (that is, $\sin\theta_i \propto \sin\theta_r$). [3]

(d) Determine a value for the refractive index of the glass block. [3]

What is being asked

This is an experimental analysis question set around a specified practical. It is assumed that the candidates will have experience of undertaking it. Part (a) is set to get the candidates to start interacting with the data. Part (b) is a standard graph question with the added twist that there is a requirement to make best use of the grid, which is slightly more than the standard demand that the points occupy more than half of each scale. Part (c) is an evaluation which appears on WJEC/Eduqas papers frequently and part (d) requires candidates to determine the relationship between the gradient of the graph and the refractive index.

Mark scheme

Question part			Description	AOs			Total	Skills	
				1	2	3		M	P
(a)			$40°$: $\sin\theta_r = 0.44$ **and** $50°$: $\sin\theta_i = 0.77$ [1] $80°$: $\theta_r = 41°$ [1] All to 2 sf [1]		3		3	3	3
(b)			Axes correctly orientated and both labelled [1] Landscape grid used with maximum scale [1] All points plotted within half a square (ignore 0) [2] (6 points plotted within half a square [1]) Line of best fit accurate – must be straight and, ideally, from the origin to slightly above the last point within half a square [1]	1	4		5		5
(c)			Straight line [1] Through the origin [1] Points are close to the line of best fit OR small scatter [1] ∴ Good agreement [needed for 3 marks]			3	3		3

(d)		Realising that 1/gradient is the refractive index [1]				3	3	3	3
		Correct method for calculating the gradient [1]							
		Correct final answer (1.49 ± 0.02) [1]							
		Alternative							
		choosing values on the line of best fit	(✓)						
		Realising $n = \dfrac{\sin\theta_i}{\sin\theta_r}$	(✓)						
		Correct final answer (1.49 ± 0.02)	(✓)						
		Max mark for choosing data point not on line = 2/3 marks							
Total				1	7	6	14	6	14

Unit 2 Practice questions

Rhodri's answers

(a) 0.44, 0.76 X
    41.3 ✓ X

MARKER NOTE

Rhodri has not rounded the second number correctly, so loses the first mark. He gains the second mark for the value but not the third because he does not express it to 2 sf.

1 mark

(b)

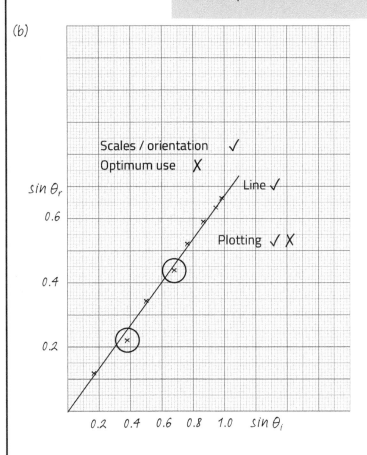

Scales / orientation ✓
Optimum use X
Line ✓
Plotting ✓ X

MARKER NOTE

Rhodri has labelled the axes correctly and they have the right orientation, so he gains the first mark.

To gain the second mark he should have used the grid landscape (as Ffion) as this makes fuller use of the grid.

He has plotted most of the points correctly but on the $\sin\theta_i$ axis has plotted 0.38 instead of 0.34 and 0.68 instead of 0.64 (circled points) so only gains one plotting mark.

The mark for the line is awarded because it is an acceptable line for the points as plotted.

3 marks

(c) Quite good agreement cos it's linear ✓
    but the data could be better cos the 2nd and
    4th points are a bit away from the line. ✓(ecf)

MARKER NOTE

Rhodri's English leaves a little to be desired but 'linear' is equivalent to straight line and he merits the first mark. The 2nd cannot be awarded 'cos' he hasn't mentioned the origin. He obtains the 3rd mark because, with ecf, his comments regarding the 2nd and 4th data points are correct.

2 marks

(d) Using $n = \dfrac{\sin i}{\sin r}$ ✓ $= \dfrac{0.34}{0.22}$ ✗

hence $n = 1.55$ ✗

MARKER NOTE

Rhodri has obtained a value by using data from the table. Unfortunately, he has chosen the point that is furthest from the line of best fit. This has led to a value of n outside the accepted limits and the only mark he can gain is the 2nd mark.

1 mark

Total	7 marks /14

Ffion's answers

(a) 0.44, 0.77 ✓

41 ✓✓

MARKER NOTE

Ffion's numbers are identical to those in the mark scheme, so full marks. Note: these are quite easy marks and one would expect most students to obtain full marks here.

3 marks

(b)

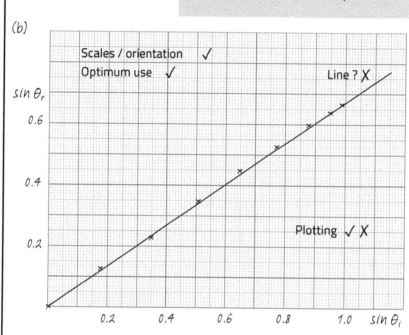

Scales / orientation ✓
Optimum use ✓
Line ? ✗
Plotting ✓ ✗

MARKER NOTE

Ffion's axes are labelled and have the correct orientation (1st mark). The graph is also the maximum size, without choosing difficult scales (2nd mark). Ffion's data points are all plotted correctly. She thus obtains the first 4 marks. The last mark is difficult to judge. Although Ffion's line is within half a square of being exact, she has 6 points above the line and only 1 point below it. The examiner has gone with ✗ but will discuss it with the senior examiner.

4 or 5 marks

(c) It's a straight line with a positive gradient ✓ The scatter about the line of best fit is quite small and is to be expected considering that the angles are only measured to the nearest degree. ∴ Good agreement ✓

MARKER NOTE

The mark that Ffion misses is the second one – she needed to mention that the best-fit line passed through the origin. She has gone further and explained the existence of scatter but unfortunately there are no marks for this on the mark scheme.

2 marks

(d) Refractive index is the gradient. ?

Gradient $= \dfrac{0.7}{1.045} = 0.766$ ✓

Hence, $n = \dfrac{1}{0.766} = 1.49$ ✓✓ (bod)

MARKER NOTE

Ffion's answer contains a contradiction. She starts by stating that the n = gradient which is incorrect. Had she stuck to this she would only have received one mark for calculating the gradient. However, she later does 1/gradient to obtain a final answer that is correct. Although there is a mistake in the first line which has not been crossed out, it is clear that she has corrected this by obtaining the final correct value. Ffion obtains the benefit of the doubt here because her final answer is correct.

3 marks

Total	12 or 13 marks /14

Section 7: Photons

Topic summary

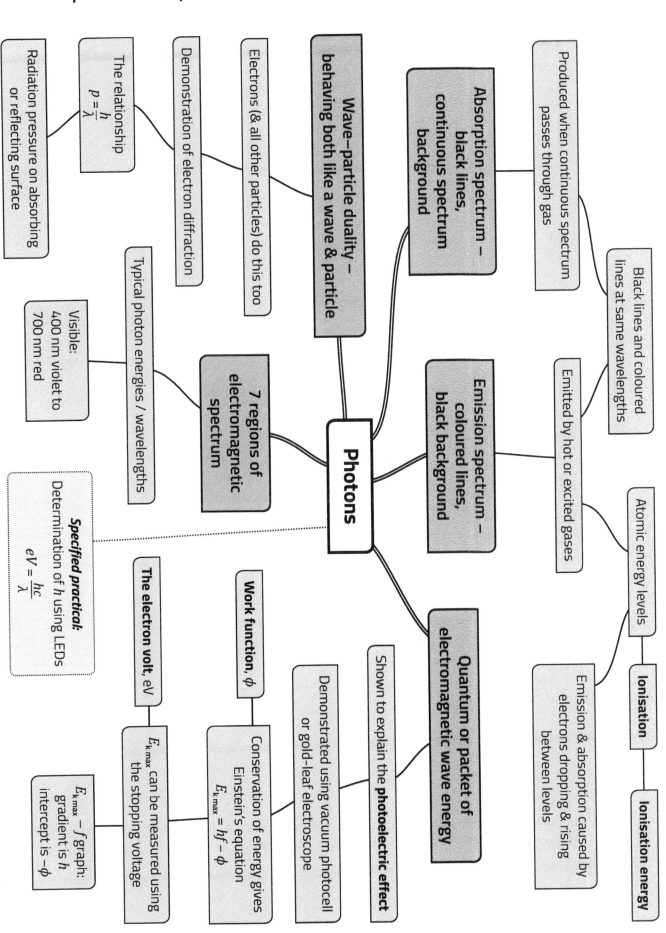

Radiation pressure on absorbing or reflecting surface

The relationship
$$p = \frac{h}{\lambda}$$

Demonstration of electron diffraction

Electrons (& all other particles) do this too

Wave–particle duality – behaving both like a wave & particle

Absorption spectrum – black lines, continuous spectrum background

Produced when continuous spectrum passes through gas

Black lines and coloured lines at same wavelengths

Emitted by hot or excited gases

Visible: 400 nm violet to 700 nm red

Typical photon energies / wavelengths

7 regions of electromagnetic spectrum

Photons

Emission spectrum – coloured lines, black background

Atomic energy levels

Emission & absorption caused by electrons dropping & rising between levels

Ionisation

Ionisation energy

Specified practical: Determination of h using LEDs
$$eV = \frac{hc}{\lambda}$$

The electron volt, eV

$E_{k\,max}$ can be measured using the stopping voltage

Work function, ϕ

Conservation of energy gives Einstein's equation
$$E_{k\,max} = hf - \phi$$

Demonstrated using vacuum photocell or gold-leaf electroscope

Shown to explain the **photoelectric effect**

Quantum or packet of electromagnetic wave energy

$E_{k\,max} - f$ graph: gradient is h intercept is $-\phi$

Q1 A monochromatic beam of light has frequency, f, and power, P. It can also be described in terms of *photons*.

(a) State briefly what is meant by a photon. [1]

....

(b) Show how the power of the beam (i.e. the energy transferred per second) relates to the frequency and number of photons per second. [2]

....

....

Q2 Describe the appearance of a line emission spectrum, the physical situation which can give rise to it and how it can be displayed. [4]

....

....

....

....

....

Q3 Explain how an absorption spectrum is produced in the atmosphere of the Sun. [4]

....

....

....

....

....

....

Q4 (a) Calculate the de Broglie wavelength of an electron accelerated by a pd of 2400 V. [4]

....

....

....

....

....

(b) An electron diffraction experiment produces a diffraction pattern of concentric rings. Explain what happens to this pattern as the accelerating pd is increased. [2]

....

....

....

Q5 The following diagram shows some of the energy levels of atomic hydrogen.

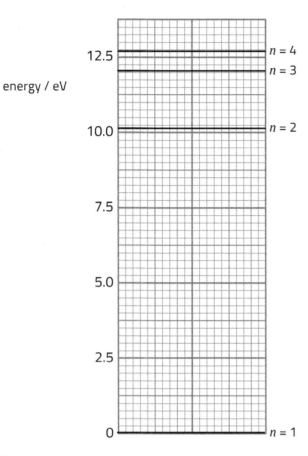

(a) (i) Calculate the energy of a photon, in eV, that is emitted when an electron drops from energy level $n = 3$ to energy level $n = 2$. Give a reason for your answer. [2]

....

....

....

(ii) Calculate the wavelength of this photon and state in which region of the electromagnetic spectrum it belongs. [2]

....

....

....

(b) (i) Calculate the highest energy, in J, of a photon that can be absorbed from a transition between the energy levels shown in the diagram. [2]

....

....

....

(ii) State in which region of the electromagnetic spectrum this photon resides. [1]

....

Q6 (a) Explain Einstein's photoelectric equation, its importance historically and how the apparatus below can be used to obtain the graph of results (also shown). [6QER]

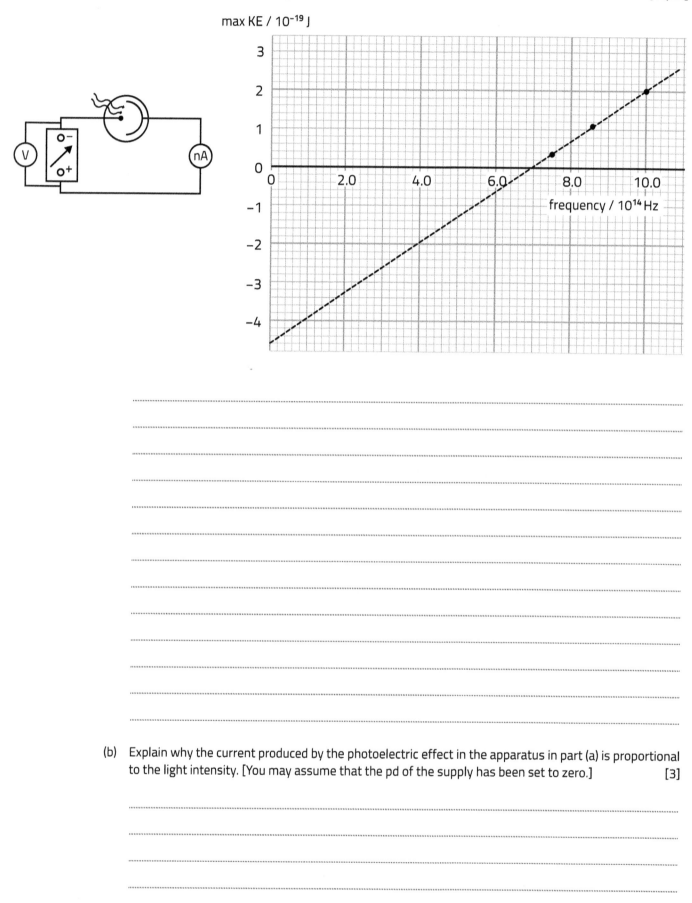

....

....

....

....

....

....

....

....

....

....

....

....

(b) Explain why the current produced by the photoelectric effect in the apparatus in part (a) is proportional to the light intensity. [You may assume that the pd of the supply has been set to zero.] [3]

....

....

....

....

Q7 The use of lasers to drive unpowered spacecraft has been the subject of research by scientists and speculation by science fiction writers. This question looks at the principles involved. Gravitational forces can be neglected in your answers.

A space probe without a rocket engine is sent to Mars. It is launched from the International Space Station and accelerated using a 15 kW laser which is located on Earth. The thrust on the probe is produced by a mirror which reflects the laser light incident on the probe.

(a) Explain briefly why reflecting the laser light can produce a force on the probe. [2]

....

....

....

(b) (i) Show that the momentum, p, of a photon is related to its energy, E, by the relationship:
$$E = pc$$
where c is the speed of light. [2]

....

....

....

(ii) The mass of the probe is 2.3 kg and all the laser light is reflected directly back continuously by the probe. Calculate the acceleration of the probe (it may be beneficial to use the relationship in (b)(i)). [3]

....

....

....

....

....

(iii) After 1 year, calculate:

(I) the speed of the probe, [2]

....

....

....

(II) the distance travelled by the probe. [2]

....

....

....

(iv) A mirror was placed on Mars by a previous mission. Explain how this might be used to decelerate the probe. [2]

....

....

....

....

(c) One problem with using a laser in this way arises from the fact that light has wave properties as well as particle properties. The laser beam emerges from a circular window, so diffraction occurs and the beam diverges as shown in the diagram:

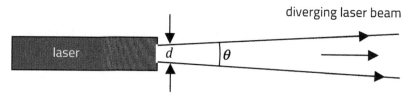

For a circular aperture, the angle of divergence, θ, is given, in radians, by:

$$\theta = \frac{2\lambda}{d}$$

where d is the diameter of the window.

(i) A lecturer uses a green laser pointer with a wavelength of 500 nm in a lecture theatre of length 10 m. Using a sensible estimate of d show that diffraction of the laser beam is unlikely to be a problem. [4]

....

....

....

....

....

....

....

(ii) In an attempt to make the green space laser as effective as possible it has a window of diameter 1 m. The spacecraft has a receiving surface of diameter 100 m. Estimate how far the spacecraft travels before the driving force is 10% of the maximum. Comment on your answer. [4]

....

....

....

....

....

....

....

....

Question and mock answer analysis

Q&A 1   The following circuit is used to find the pd across a LED when it is switched on:

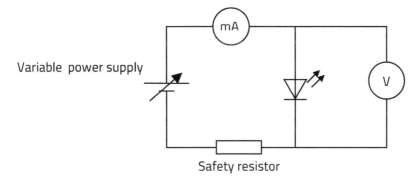

The LED is judged to be switched on when a current of 10.0 mA passes through it. The variable power supply is adjusted and the 'switching on pd', V_s, recorded when the current reaches 10.0 mA. The procedure is repeated for a range of different LEDs which emit light of different wavelengths, λ. The incomplete results and plot of them are shown below:

λ / nm	$\frac{1}{\lambda}$ / 10^6 m^{-1}	V_s / V
480	2.08	2.65
590		2.14
680	1.47	1.85
895		1.40
930	1.08	1.34

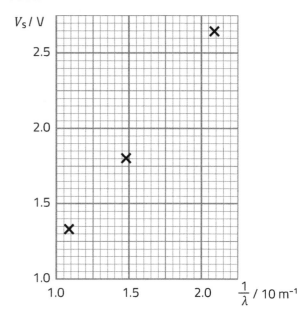

(a)   Complete the table. [2]

(b)   Complete the graph by plotting the two missing points whose values you have just calculated and drawing an appropriate line. [3]

(c)   The application of the principle of conservation of energy to an electron and a photon involved in the light-emitting process of the LED gives:
$$eV = \frac{hc}{\lambda}$$

(i)   Use the graph to determine a value for the Planck constant to an appropriate number of significant figures. [4]

(ii)   Explain to what extent the data confirm the relationship, $eV = \frac{hc}{\lambda}$. [5]

What is being asked?

This is a specified practical, with which you should be familiar. However, there is no need to recall any aspect of the practical work as all the questions involve interaction with the data and forming conclusions, i.e. the entire question is AO2 and AO3. The question style is quite a familiar one in the AS examination. It also reappears in the Practical Analysis paper in Unit 5 of the A2 examination, with students being asked to perform calculations on experimental data, measure gradients and intercepts and draw conclusions about whether the data agree with theoretical ideas.

Mark scheme

Question part		Description	AOs 1	AOs 2	AOs 3	Total	Skills M	Skills P
(a)		1.69 [1] 1.12 [1]		2		2	2	2
(b)		(1.69, 2.14) plotted to within ½ square in each direction – ecf [1] (1.12, 2.40) plotted to within ½ square in each direction – ecf [1] Good fit straight line drawn (by eye) to plotted points [1]			3	3		3
(c)	(i)	Correct method for calculating gradient, i.e. $\Delta y / \Delta x$ attempted. [1] Correct gradient (ecf on line) – expect 1.3 [∴ 10^6] – ignore powers of 10. [1] Statement that gradient $= \dfrac{hc}{e}$ [or by implication] [1] Correct answer with 2 or 3 sf only (expect $6.9[0] \times 10^{-34}$) [1] [If data point from the line or table is used, max 2 marks allowed if 2 or 3 sf given]			4	4	4	4
	(ii)	[The line of best fit] is a straight line [1] Intercept on V_s calculated [1] Comment that intercept is very close to origin [1] Very little scatter in data points **or** points close to line of best fit. [1] The value of the Planck constant is close to the accepted [accept: actual] value. [1]			5	5	1	5
Total			0	5	9	14	7	14

Rhodri's answers

(a) 1.69 ✓

  1.1173 ✗ [rounding]

MARKER NOTE

Rhodri has calculated $1/\lambda$ correctly. He should have rounded up the second answer [$1.1173 \rightarrow 1.12$ to 2 sf]. He gets the mark for the first answer.

1 mark

(b)

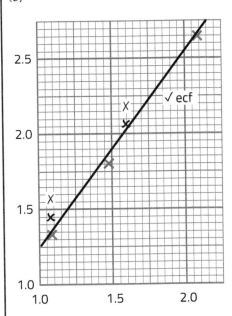

✓ ecf

MARKER NOTE

Rhodri has mis-plotted both his data points. He has misinterpreted the scale, thinking that each small square is 0.1 on both axes, whereas in fact two squares are 0.1. So no plotting marks. The poor plotting makes it more difficult to judge the best-fit line but it is a reasonable attempt and gains the line mark.

1 mark

MARKER NOTE

Rhodri has obtained a good value for the Planck constant but he has used a single data point rather than the gradient. This is especially wrong because the line doesn't quite pass through the origin, possible because of a systematic error. He should have worked to a maximum of 3 sf because the data are all to 3 sf, but he has given his answer to 4 sf. A 2 or 3 sf answer, i.e. 6.7 or 6.70, would have gained him two marks but unfortunately he loses one of these and only obtains one.

1 mark

(c) (i) Using the middle point: $\lambda = 680$ nm and $V_s = 1.85$ ✗

  Rearrange equation: $h = \dfrac{eV\lambda}{c}$ ✓

  $h = \dfrac{1.6 \times 10^{-19} \times 1.85 \times 10^{-9}}{3 \times 10^{8}}$ ✗

  $h = 6.709 \times 10^{-34}$ ✗

(ii) The graph is a straight line ✓ with a positive gradient but it doesn't pass through the origin. ✗ My value for Planck's constant is good. ✓

MARKER NOTE

Rhodri's evaluation is extremely succinct but, nonetheless, it hits two of the marking points: the straight line and the Planck constant value. He has not calculated the intercept on the V_s axis, so really has no idea whether it is zero or not – he has probably not noticed that neither axis scale goes back to zero. No additional credit is given for the positive gradient.

2 marks

Total   **5 marks /14**

Unit 2 Practice questions

Ffion's answers

(a) 1.70 X [rounding]

   1.12 ✓

(b)

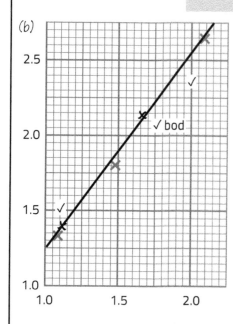

(c) (i) Gradient = $\dfrac{2.75 - 1.23}{(2.17 - 1.00) \times 10^6}$

   = 1.299×10^{-6} ✓

   But gradient = $\dfrac{hc}{e}$ ✓

   So $h = \dfrac{e}{c} \times$ gradient = 6.93×10^{-34} J s ✓

   (ii) The line of best fit is a staight line ✓ which passes through all the data points ✓

   When the x value is 0 the y value is:
   $1.23 - (1.0 \times 10^6) \times 1.30 \times 10^{-6} = -0.07$. ✓

   Theory predicts the graph should go through zero, so this is not quite right but quite close. ✓
   The value obtained for the Planck constant is close to the accepted value of 6.63×10^{-34} Js (only 4% out). ✓

Total **13 marks /14**

Section 8: Lasers

Topic summary

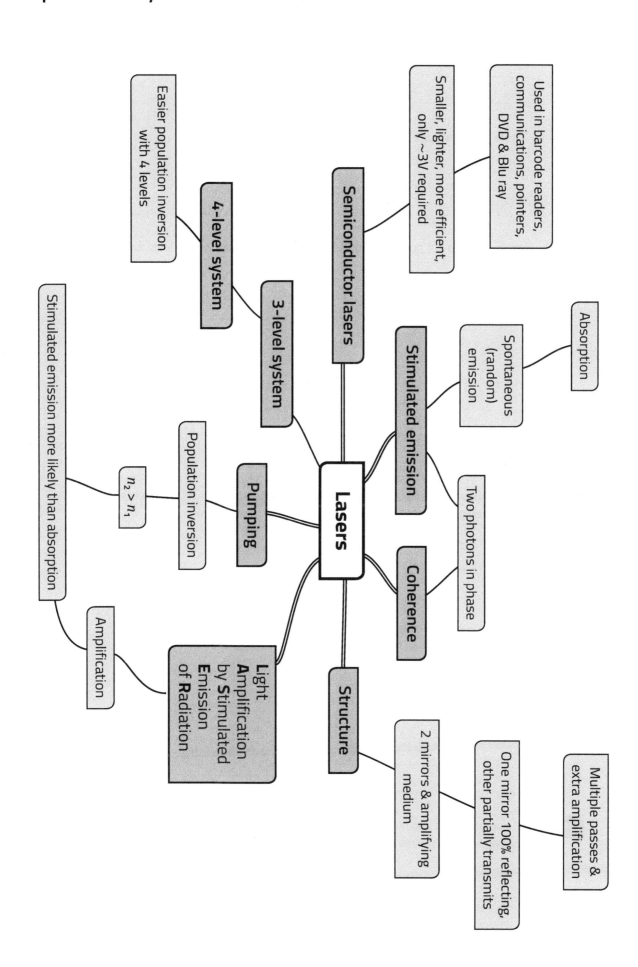

Q1 Laser is an acronym for Light Amplification by Stimulated Emission of Radiation. Explain what is meant by the term *stimulated emission*. [3]

....

....

....

....

Q2 (a) State what is meant by the term *population inversion*. [1]

....

....

(b) Explain why it is not possible to obtain a *population inversion* in a 2-level system when optical pumping is used. [4]

....

....

....

....

....

....

....

Q3 Explain why a population inversion is harder to obtain in a 3-level laser system than a 4-level laser system. You should add to the diagrams as part of your answer. [4]

3-level

4-level

....

....

....

....

Q4 The pumping level (top level) in laser systems must have a short lifetime. State two reasons why the pumping level must have a short lifetime. [2]

....

....

....

Q5 State two uses and two advantages of semiconductor lasers. [2]

....

....

....

Q6 The energy levels of a 4-level laser system are shown:

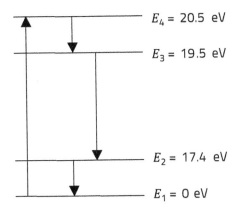

$E_4 = 20.5$ eV

$E_3 = 19.5$ eV

$E_2 = 17.4$ eV

$E_1 = 0$ eV

(a) Calculate the frequency of laser emission. [3]

....

....

....

....

....

(b) Laser emission occurs when an electron is stimulated to drop from the 3rd energy level. State what is responsible for stimulating this emission and explain why stimulated emission is far more probable than spontaneous emission for a laser. [3]

....

....

....

....

(c) Joel claims that this particular laser system could never be more than 10% efficient. Evaluate to what extent Joel is correct. [2]

....

....

....

(d) Nigella states, 'The 2nd energy level (E_2) in a 4-level laser system must have a long lifetime.' Evaluate to what extent Nigella's statement is correct. [2]

....

....

....

Unit 2 Practice questions

Q7 The energy levels of a 3-level laser system are shown.

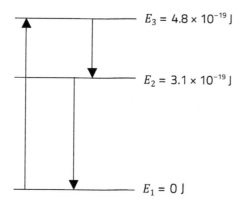

$E_3 = 4.8 \times 10^{-19}$ J

$E_2 = 3.1 \times 10^{-19}$ J

$E_1 = 0$ J

(a) Calculate the energy of the photons used to pump the laser in eV and state what colour they are. [2]

....

....

....

(b) Compare the lifetimes of the two excited states in this system and give a reason for your answer. [2]

....

....

....

(c) Calculate the wavelength of laser emission. [2]

....

....

....

(d) Two photons travel together after stimulated emission occurs. State three distinct properties that are common to these two photons. [3]

....

....

....

(e) Paula states 'More than half the electrons of the ground state must be pumped in order to achieve population inversion in a 3-level laser system.' Evaluate to what extent Paula's statement is correct. [3]

....

....

....

....

....

Q8 A diagram of a laser cavity is shown below:

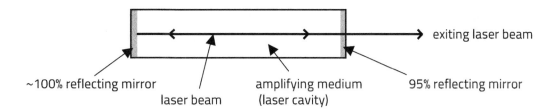

~100% reflecting mirror   laser beam   amplifying medium (laser cavity)   95% reflecting mirror   exiting laser beam

(a) In a laser, the left mirror cannot be made to be exactly 100% reflecting. Explain why the left mirror should, ideally, be 100% reflecting. [2]

(b) Explain why the right mirror must not be 100% reflecting. [1]

(c) (i) Explain briefly why the force exerted by the laser light on the left mirror is greater than the force exerted on the right mirror. [2]

(ii) The output power of the laser is 5 mW and the cross-sectional area of the laser beam is 0.94 mm². Calculate the pressure exerted on the right mirror (you may assume that the wavelength of the laser light is 500 nm although this problem is soluble without this information). [3]

(d) Helena claims that the laser must increase in intensity by approximately 3% each time it traverses a length of the laser cavity (when the laser is in equilibrium). Evaluate whether or not Helena's claim is correct. [3]

Unit 2 Practice questions

Question and mock answer analysis

Q&A 1

(a) Explain how a 3-level laser system works **and** how the construction of the laser cavity and mirrors assists in producing a laser beam (you may refer to the diagrams in your answer). [6 QER]

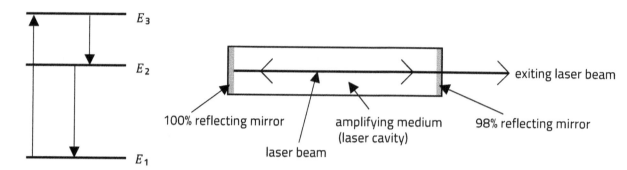

(b) Light of wavelength 950 nm from a powerful 50 W laser is focused from a beam diameter of 1.2 mm to a beam diameter equal to one wavelength (950 nm). Calculate the final intensity of the laser beam.

[3]

What is being asked

(a) The specification has a series of statements about laser function, all of which are addressed in this question. The statements are:
- the process of stimulated emission...
- the idea that a population inversion ($N_2 > N_1$) is necessary for a laser to operate
- how a population inversion is attained in 3-level energy systems
- the process of pumping and its purpose
- the structure of a typical laser, i.e. an amplifying medium between two mirrors, one of which partially transmits light.

This question asks you to pull together these statements in order to write an account of laser operation. This is a very large topic but the examiner has used the diagrams to provide a framework for this AO1 piece of bookwork. Notice that there are two broad parts to the question. You are expected to tackle both – hence the emboldened **and** in the question.

The marking scheme has a series of points which you could choose to make. You are not expected to mention all of them but you need to use several of them from both sections (energy levels and structure) in order to be put into the top band.

(b) The second part of the question is an AO2 calculation and will require the application of conservation of energy and an understanding of the term intensity. This question is synoptic because intensity normally appears with the topic of stars in Unit 1. It has an added level of difficulty because the definition of intensity in the Terms & Definitions booklet does not quite cover this question – that definition is aimed at stars and the inverse square law.

Mark scheme

Question part		Description	AOs			Total	Skills	
			1	2	3		M	P
(a)		**Points regarding energy levels** 1. Pumping is from E_1 to E_3 2. E_3 has a short lifetime 3. E_2 is metastable / has a long lifetime 4. E_1 is the ground state 5. Ground state is usually full 6. Population inversion between E_2 and E_1 7. Stimulated emission between E_2 and E_1 8. Greater than 50% pumping needed **Points regarding mirrors and structure** 1. Amplification while passing through cavity (exponential growth) 2. 100% mirror returns beam to cavity 3. 98% mirror allows 2% through (but returns/reflects nearly all) 4. Multiple passes (~50) through cavity 5. Equilibrium when overall losses through the 98% mirror – overall gains due to amplification 6. Parallel mirrors give alignment of beam 7. Stationary wave between mirrors	6			6		
(b)		Realising power will still be 50 W [1] Application of $I = \dfrac{P}{\lambda r^2}$ or $I = \dfrac{P}{\text{area}}$ [1] Correct answer – 70.5 TW m^{-2} [1]		3		3	1 1	
Total			**6**	**3**		**9**	**2**	

Rhodri's answers

(a) The transition from E_1 to E_3 is called pumping. E_3 is a broad energy level with a very <u>short lifetime</u> to help with pumping and keeping E_3 empty. E_2 has a long lifetime so that a <u>population inversion can be obtained between E_2 and E_1</u>. It is difficult to obtain a population inversion because <u>E_1 is the ground state</u> and is <u>normally full</u>. Hence, <u>more than half</u> the electrons must be pumped in order to obtain a population inversion. <u>Stimulated emission will then be more likely than absorption between E_2 and E_1</u> and the laser beam will gain in intensity.

MARKER NOTE

Rhodri's answer is an excellent response to the first part of the question. However, he has forgotten to answer the second part of the question even though the 'and' has been emboldened in the question. This happens surprisingly often even with excellent candidates. Note that the marker has underlined Rhodri's points and that Rhodri has succeeded in hitting all the 8 energy level points in the MS. Rhodri has gone even further and has provided three extra pieces of information. First, E_3 is a broad level – true this helps to increase the probability of pumping. Second, he has correctly pointed out that E_3 should remain empty. Third, Rhodri has stated that stimulated emission is more probable than absorption, an excellent point not on the mark scheme. All this means that Rhodri has a middle band answer – an excellent response to only one of the two answer parts. However, the quality and clarity are such that 4 marks are deserved rather than 3 marks.

4 marks

(b) The intensity will still be 50 W X because you won't lose any of the power of the laser. ✓

MARKER NOTE

Rhodri appears to have no concept of intensity. Nonetheless, he gains one mark because he has applied his knowledge of conservation of energy well.

1 mark

Total   **5 marks /9**

Ffion's answers

(a) The rise from E_1 to E_3 is called pumping and is often done by accelerated electrons or light absorption. E_3 must have a short lifetime so that the electrons go to E_2 quickly. Metastable level is the name for energy level E_2 and it has a long half-life. This means that a population inversion is possible between E_2 and E_1 leading to lots of stimulated emission and amplification. Amplification by traversing the laser cavity once is not usually enough and this is where the design of the laser system and mirrors is important. One mirror is 100% reflecting and this ensures that the beam returns to traverse the amplifying medium again, leading to greater amplification. The other mirror allows 2% of the light to be transmitted and this is the output beam. Only 2% of the light is output each time which means that photons traverse the amplifying medium 50 times on average before exiting.

(b) Intensity $- \dfrac{power}{area} = \dfrac{50 \ W}{4\pi \ r^2}$ ✓ [application]

$= \dfrac{50 \ W}{4\pi \left(950 \times 10^{-9} \ m\right)^2}$ ✓

$= 4.4 \times 10^{12} \ W \ m^{-2}$

MARKER NOTE

Ffion's answer has 4½ good points from the energy levels part. The marker has underlined the 4 points that are not in doubt. Ffion has also mentioned stimulated emission but has not stated explicitly between which levels this occurs although the stimulated emission has been linked to the population inversion. Also, Ffion has added a little extra explanation that the short half-life of E_3 allows electrons to pass quickly to E_2. Ffion's answer to the second part of the question also makes four good points (the first four points). Although Ffion has not hit any of the last 3 points on the 'mirrors & structure' part, these are more advanced points. Overall, Ffion's response to both parts is just about good enough to place her in the top band. Her use of English is good and the sentence linking both parts of the answer is a particularly nice touch. Final mark – 5 or 6 depending on whether the marker feels that rewarding the well-constructed answer is more important than penalising the omission of reference to the ground state.

5 or 6 marks

MARKER NOTE

Ffion does not mention that the power of the beam will still be 50 W but it is implied in her calculation and she receives this first mark. She realises that the initial area of the beam and the wavelength are unnecessary pieces of information.

She has attempted power / area to obtain the intensity and thus gained the second mark, but has made two separate mistakes. First, she has used the wrong equation for the area – she needs the cross-section of a circle not a sphere. Second, she has used the diameter of the beam instead of the radius of the beam. She is probably a little fortunate to receive 2 marks but this type of answer would have been discussed in the examiners' meeting and this mark would have been agreed at that stage.

2 marks

Total	
	7 or 8 marks /9

Practice papers

AS PHYSICS
UNIT 1 PRACTICE PAPER

1 hour 30 min

For Examiner's use only		
Question	Maximum Mark	Mark Awarded
1.	16	
2.	9	
3.	8	
4.	10	
5.	6	
6.	11	
7.	8	
8.	12	
Total	80	

Notes

In a WJEC paper, the following information will be given on the front of the paper:

1. **Additional materials**
    You will be told that you will require a calculator and a **Data Booklet**. Sometimes you will be told that you need a ruler and/or an angle measurer / protractor.

2. **Answering the examination**
    You will be told to use a blue or black ball-point (but graphs are best drawn using a pencil).
    You will be told to answer **all** the questions in the spaces provided on the question paper.

3. **Further information**
    Each question part shows, using square brackets, the total marks available. One question will assess the quality of extended response [QER]. This question will be identified on the front page. In this practice paper the QER question is question **5**.

*Answer **all** questions.*

1. (a) A metal rod is put under a tensile stress. Explain what is meant by tensile stress, adding to the diagram.

    [2]

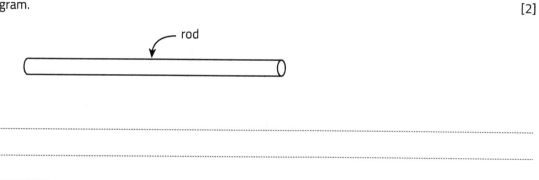

    rod

    ....

    ....

    ....

    (b) Aled investigated the stretching of a brass wire using the apparatus shown:

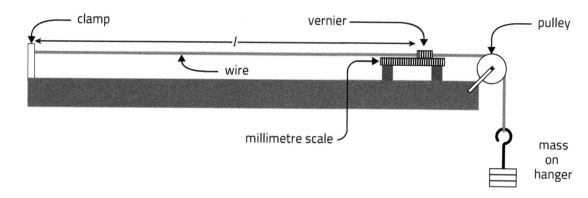

    clamp   vernier   pulley

    l

    wire

    millimetre scale   mass on hanger

    Aled placed increasing known masses on to the hanger and measured the extension of the wire using the vernier scale. The extension was taken as zero with just the hanger attached to the wire. Aled's plot of mass added against extension is given below.

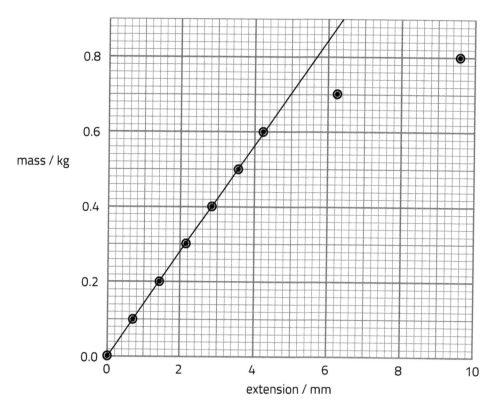

    mass / kg

    extension / mm

(i)   Aled measured the **diameter** of the wire to be 0.20 mm, and its effective length to be 2.500 m. Suggest how he obtained this value for the diameter and give a reason for your answer.   [2]

....

....

....

(ii) Use the graph and the values given in (b)(i) to obtain a value for the Young modulus of brass. Give your answer to an appropriate number of significant figures.   [5]

....

....

....

....

....

....

....

....

(iii) Aled believed that for the two largest masses used the wire was stretching *inelastically*.

(I)   Suggest how he could have confirmed that inelastic stretching had occurred.   [2]

....

....

....

(II) Explain briefly, on an atomic level, how a metal deforms inelastically.   [2]

....

....

....

(c) In the 1950s, physicists confirmed that it was the movement of *dislocations* that allowed inelastic deformation of metals to take place at relatively low **stresses**. Suggest how we have benefitted practically from the application of this knowledge.   [3]

....

....

....

....

....

....

2. A simple bridge over a stream consists of a uniform plank of mass 44 kg and length 2.4 m, hinged to a vertical post and kept horizontal by a rope joining the end of the plank to the post.

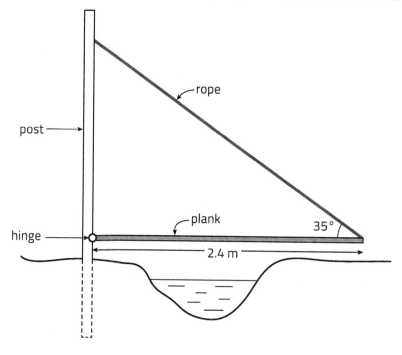

(a) (i)   Calculate the moment about the hinge of the plank's weight.   [2]

....

....

....

(ii)   Hence show that the tension in the rope is approximately 400 N.   [2]

....

....

....

(iii) Calculate the horizontal component of the force that the hinge exerts on the plank, explaining your reasoning.   [2]

....

....

....

(b) The makers of the rope state that the maximum safe tension is 1500 N. Evaluate whether it would be safe for a student of mass 68 kg to walk across the plank, starting at the post.   [3]

....

....

....

....

....

3. Osian demonstrates the acceleration produced by a resultant force using the apparatus shown. The trolley is released from rest at time $t = 0$.

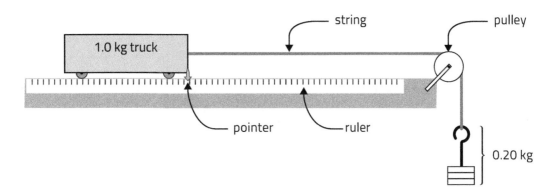

(a) (i) Show clearly that the expected acceleration of the trolley is approximately 1.6 m s^{-2}. [2]

....

....

....

(ii) Explain why the tension in the string must be less than the weight of the 0.20 kg mass. [2]

....

....

....

(b) The trolley is fitted with a pointer that can be seen against a ruler fixed next to the path of the trolley. The following measurements are made from three pictures taken.

Time of picture (from trolley release) / s	0.000	0.150	0.300
Position of pointer against ruler / cm	10.00	11.8	17.3

Evaluate whether or not these measurements are consistent with the trolley having a constant acceleration of the expected value – see part (a). [4]

....

....

....

....

....

....

....

....

4. (a) State the *principle of conservation of momentum*.   [2]

....

....

....

(b) In a friendly contest, competitors kick a ball of mass 0.42 kg at a wooden block of mass 5.0 kg that is initially stationary and resting on flat ground. The competitors are trying to make the block slide as far as possible.

0.42 kg

The winner kicks the ball at a velocity of 28.0 m s^{-1} straight at the block. The ball bounces back from the block at 9.0 m s^{-1}.

(i) Show that the kinetic energy acquired by the block is approximately 24 J, stating your assumption.   [4]

....

....

....

....

....

....

....

....

....

(ii) Determine whether or not the collision between ball and block is elastic.   [2]

....

....

....

(iii) The block slides 3.6 m before coming to rest. Calculate the mean frictional force on the block from the ground.   [2]

....

....

....

5. Two balls have the same size and shape but different masses. They are dropped from a very high platform. Explain, in terms of forces, why each eventually approaches a terminal velocity, but the two terminal velocities are different. [QER 6]

6. (a) A cyclist cycles on a circular track of radius 60 m at a constant speed, taking 15 s to go half way round, from A to B.

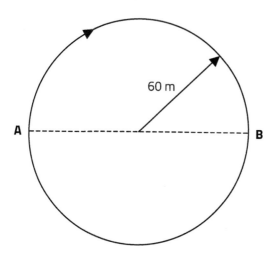

(i) Determine:

(I) her speed. [2]

....

....

....

(II) her mean velocity as she goes from A to B. [2]

....

....

....

(ii) State why the cyclist has a mean acceleration in going from A to B. [1]

....

....

(b) A ball is thrown from near ground level with a velocity of 30 m s^{-1} vertically upwards.

(i) Sketch a velocity–time graph for the complete flight of the ball on the axes provided, marking significant times and velocities. The space underneath is for calculations. [4]

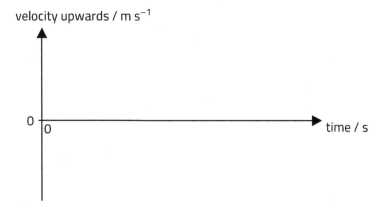

velocity upwards / m s^{-1}

time / s

(ii) Calculate the distance gone over the complete flight. [2]

7. The continuous spectrum of the bright star Canopus is given.

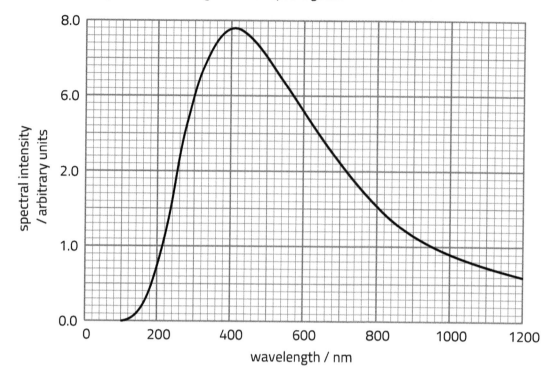

(a) Show that the temperature of Canopus is approximately 7000 K. [2]

....

....

....

(b) Determine the ratio $\dfrac{\text{radius of Canopus}}{\text{radius of sun}}$

given that $\dfrac{\text{luminosity of Canopus}}{\text{luminosity of sun}} = 10\,700$ and temperature of Sun = 5 780 K [3]

....

....

....

....

....

(c) Megan claims that Canopus must be a red giant. Evaluate this claim. [3]

....

....

....

....

....

8. (a) Define in terms of quarks:

(i) A baryon. .... [1]

(ii) A meson. .... [1]

(b) When two protons with high kinetic energy collide, the following interaction may occur:

$$p + p \longrightarrow p + x + \pi^+$$

(i) Use two conservation laws *that apply to all interactions* to identify the (first-generation) particle x, explaining how you are using each law. [3]

....

....

....

....

(ii) Show whether or not u and d quark numbers are **separately** conserved in the interaction. [2]

....

....

....

(iii) State, giving a reason, which force (strong, weak or electromagnetic) is responsible for this interaction. [1]

....

....

....

(c) Another interaction that occasionally occurs between protons is:

$$p + p \longrightarrow {}^{2}_{1}H + z + e^+$$

(i) Explain why particle z must be a neutrino, v_e. [2]

....

....

....

(ii) The emission of a neutrino allows us to conclude that this is a weak interaction. Explain how else we can reach this conclusion. [2]

....

....

....

END OF PAPER

AS PHYSICS
UNIT 2 PRACTICE PAPER

1 hour 30 min

For Examiner's use only		
Question	Maximum Mark	Mark Awarded
1.	11	
2.	14	
3.	11	
4.	8	
5.	9	
6.	9	
7.	10	
8.	8	
Total	80	

Notes

In a WJEC paper, the following information will be given on the front of the paper:

1. **Additional materials**
   You will be told that you will require a calculator and a **Data Booklet**. Sometimes you will be told that you need a ruler and/or an angle measurer / protractor.

2. **Answering the examination**
   You will be told to use a blue or black ball-point (but graphs are best drawn using a pencil).
   You will be told to answer **all** the questions in the spaces provided on the question paper.

3. **Further information**
   Each question part shows, using square brackets, the total marks available. One question will assess the quality of extended response [QER]. This question will be identified on the front page. In this practice paper the QER question is question **1(b)**.

Answer **all** questions.

1. (a) Define electric current. [1]

....

....

....

....

(b) Explain the process of conduction of electricity in metals and how an increase in temperature affects resistance. [6QER]

....

....

....

....

....

....

....

....

....

....

....

....

....

(c) (i) A metal wire has cross-sectional area 2.75×10^{-5} m², resistivity 5.60×10^{-8} Ω m and length 65.0 cm. Calculate its resistance. [2]

(ii) The number of free electrons in the metal wire per unit volume is 8.25×10^{28} m⁻³. Calculate the drift velocity when a p.d. of 1.65 V is applied along the length of the wire. [2]

2. Archibald carries out an experiment to obtain the *I–V* characteristics of a filament lamp. He uses the circuit shown, his results are shown in the table and the graph is also plotted

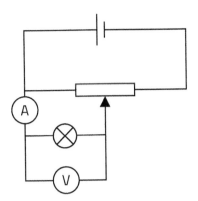

pd / V	Current / mA
0.00	0
1.90	90
4.05	167
6.07	205
7.93	226
9.91	240
12.06	251

(a) Plot the two missing points on the grid and draw a line of best fit. [2]

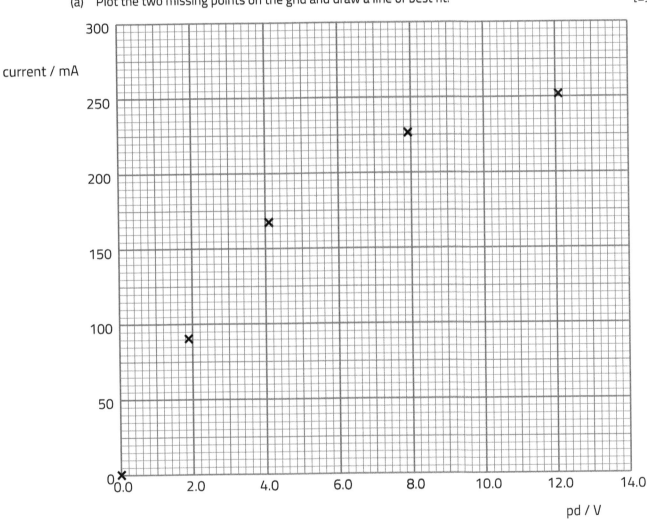

current / mA

pd / V

(b) Calculate the resistance of the filament lamp:

(i) At room temperature. [2]

....

....

....

(ii) When the pd is 12 V. [1]

....

....

(c) (i) Explain to what extent the results agree with those expected for a metal filament lamp. [4]

....

....

....

....

....

(ii) Look carefully at the circuit diagram and the table of results and suggest why repeat readings are inappropriate for this experiment. [2]

....

....

....

(d) (i) Suggest whether or not obtaining I–V characteristics for a superconductor below its transition temperature is possible. [2]

....

....

....

(ii) State a use of a superconductor. [1]

....

3.   A cell of emf E and internal resistance r is used to provide current, I, to an external resistance R.

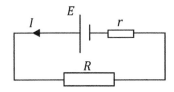

(a)   Explain the equation:

$$E = IR + Ir$$

in terms of energy. [3]

....

....

....

....

(b)   The emf of the cell is 1.65 V, the internal resistance of the cell is 0.20 Ω and the external resistance is 0.20 Ω.

(i)   Calculate the current, I. [2]

....

....

....

....

(ii)   Show that the power dissipated in the external resistance is approximately 3.40 W. Ensure that your answer is to at least 4 sf. [2]

....

....

....

(c)   Meinir claims that this power dissipation of approximately 3.40 W is the maximum power dissipation that can be achieved in any external resistance using the 1.65 V cell with internal resistance 0.20 Ω. Determine whether or not Meinir appears to be correct. [4]

....

....

....

....

....

....

4. (a) State the difference between a transverse wave and a longitudinal wave. [2]

....

....

....

....

(b) A displacement–time graph is shown for a transverse wave:

displacement / cm

Determine:

(i) the period of the wave. [1]

....

....

(ii) the frequency of the wave. [1]

....

....

(c)   A displacement–distance graph is shown for the same wave:

displacement / cm

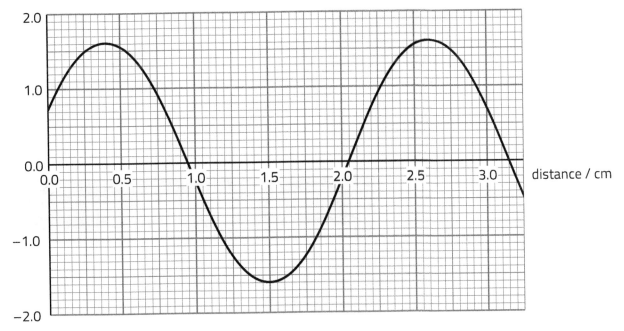

(i)   Calculate the wave speed.   [2]

....

....

....

....

....

(ii)   On the grid above, draw the displacement–distance graph for the wave 0.050 s later (the wave is
        travelling to the right).   [2]

5. Young's slits experiment is carried out using the following set up:

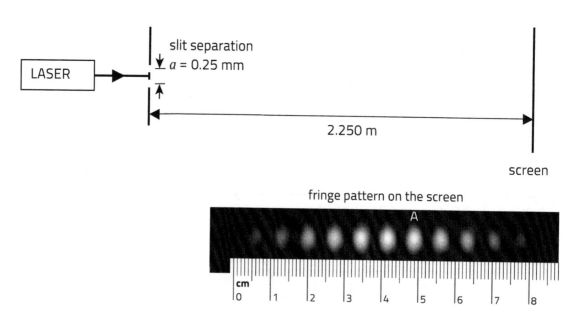

(a) Although lasers were not available to Thomas Young, state the historical importance of Young's double slit experiment. [1]

....

....

(b) Use the information in the diagrams to calculate the wavelength of the laser light. [3]

....

....

....

....

(c) Explain in careful steps how the bright fringe labelled **A** (next to the central fringe) in the diagram arises. Assume that the slits give out light in phase with each other. [3]

....

....

....

....

(d) A student replaces the double slit with a diffraction grating which has a slit separation, d, exactly the same as the separation of the double slits, i.e. 0.25 mm. Sketch the new fringe pattern in the empty box below. [2]

fringe pattern due to double slit

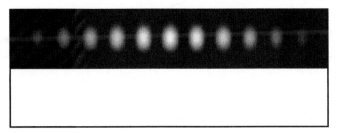

fringe pattern due to diffraction grating

6. A light ray enters an optical fibre as is shown in the diagram:

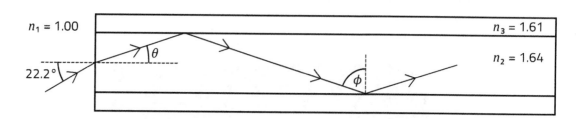

(a) Show that the angle θ is approximately 13°. [2]

....

....

....

(b) Determine whether or not this light will propagate along the length of the optical fibre with total internal reflection. [4]

....

....

....

....

....

....

....

....

(c) Optical fibres are usually made from silica (SiO_2) which is, essentially, purified sand. However, many modern optical fibres use plastics which are a by-product of the oil industry. Discuss which type of optical fibre is worse for the environment. [3]

....

....

....

....

....

....

7. The energy levels of a 3-level and 4-level laser system are shown below:

1.60 eV ————————————————

2.60 eV ————————————————

1.42 eV ————————————————

1.79 eV ————————————————

0.25 eV ————————————————

0.00 eV ————————————————

0.00 eV ————————————————

3-level system

4-level system

(a) Calculate the laser emission wavelength for both laser systems.   [4]

....

....

....

....

....

....

....

(b) State why population inversion is achieved more easily in a 4-level than a 3-level system.   [3]

....

....

....

....

....

(c) Explain briefly the purpose of the mirrors in the laser system below.   [3]

→ exiting laser beam

100% reflecting mirror   amplifying medium   99% reflecting mirror
                                (laser cavity)

laser beam

....

....

....

....

8. Light of various wavelengths is shone on a photocell as is shown in the diagram:

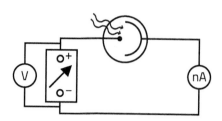

Einstein's equation for the photoelectric effect is:

$$hf = \phi + E_{k\,max}$$

(a) Explain why this equation is an application of conservation of energy.   [3]

....

....

....

....

(b) The results obtained in this experiment are plotted in the following graph:

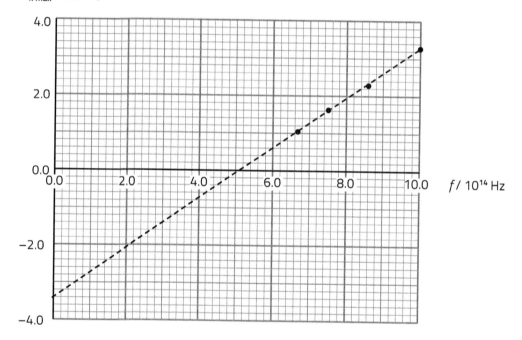

(i) Determine the work function of the metal and calculate a value for the Planck constant. [3]

....

....

....

....

(ii) Explain why the data in the graph can be considered excellent data. [2]

....

....

....

....

END OF EXAM

Answers

Practice questions: Unit 1: Motion, Energy and Matter

Section 1: Basic physics

Q1
metre (m) kilogram (kg)
second (s) mole (mol)
ampère (A) kelvin (K)

Q2 (a) newton (N)

(b) $N = kg\ m\ s^{-2}$

(c) Work = Force × distance moved in direction of force
So $J = N\ m = (kg\ m\ s^{-2})\ m = kg\ m^2\ s^{-2}$

Q3 (a) $k = \dfrac{F}{v}$, so $[k] = \dfrac{N}{m\ s^{-1}} = \dfrac{kg\ m\ s^{-2}}{m\ s^{-1}} = kg\ s^{-1}$

(b) $K = \dfrac{F}{Av^2}$, so $[K] = \dfrac{[F]}{[A]\,[v^2]} = \dfrac{kg\ m\ s^{-2}}{m^2\,(m\ s^{-1})^2} = \dfrac{kg\ m\ s^{-2}}{m^4\ s^{-2}} = kg\ m^{-3}$

This is the unit of density, so K might represent the density of the fluid through which the object is moving, perhaps with a numerical (unit-less) multiplier.

Q4 $[\pi] = \dfrac{[A]}{[r^2]} = \dfrac{m^2}{m^2} = 1$, that is, no units.

Alternatively, we could show it using the equation: circumference = $2\pi r$

Q5 Scalars: energy, time, density, temperature, pressure
Vectors: acceleration, velocity, momentum
Note: pressure is a scalar because only the *magnitude* of a force is involved in its definition:

Pressure = $\dfrac{\text{magnitude of force normal to a surface}}{\text{area of a surface}}$

Q6 $[v] = m\ s^{-1}$; $[u] = m\ s^{-1}$; $[at] = m\ s^{-2} \times s = m\ s^{-1}$

So the two terms on the right-hand side have the same units, so can be added together **and** the unit of the left-hand side is the same as the unit of the right-hand side.

Q7 (a) 28 N   45 N

resultant magnitude 73 N   Direction = that of the 45 N force

(b) 45 N

28 N
Resultant magnitude 17 N   Direction = that of the 45 N force

(c) 45 N

28 N
53 N

Resultant magnitude = $\sqrt{45^2 + 28^2} = 53$ N   Direction = $\tan^{-1}\left(\dfrac{28\ N}{45\ N}\right) = 32°$ to 45 N force

Q8 (a) $F \sin 25° = 53$ so $F = \dfrac{53}{\sin 25°} = 125$ N **or** $F \cos(90° - 25) = 53$ leading to $F = 125$ N

(b) Horizontal component = $F \cos 25° = 125$ N × $\cos 25° = 114$ N

Q9 (a) Perpendicular component = 5.0 kN × cos 75° = 1.3 kN [1.29 kN to 3 sf]

(b) 1.3 kN

(c) 1.29 kN = F × sin 10°

So $F = \dfrac{1.29 \text{ kN}}{\sin 10°}$ = 7.4 kN [7.43 kN to 3 sf]

(d) Since force components at right angles to forward direction cancel,

Resultant force = forward component of 5.0 kN + forward component of F
= 5.0 kN × cos 15° + 7.43 kN × cos 10°
= 4.83 kN + 7.32 N = 12 kN (to 2 sf)

Q10 (a) The vector sum of the forces on the object is zero.

(b) The sum of the clockwise moments about any point is equal to the sum of the anticlockwise moments about the same point.

Q11 (a) at t_1, v_{horiz} = 15.0 m s^{-1} × cos 30.0° = 13.0 m s^{-1}

v_{up} = 15.0 m s^{-1} × sin 30.0° (or cos 60.0°) = 7.5 m s^{-1}

(b) at t_2, v_{horiz} = 20.0 m s^{-1} × cos 49.5° = 13.0 m s^{-1}

v_{up} = −20.0 m s^{-1} × sin 49.5° (or cos 40.5°) = −15.2 m s^{-1}

So, Δv_{horiz} = 0

And Δv_{up} = (−15.2 m s^{-1}) − (+7.5 m s^{-1}) = −22.7 m s^{-1}

So the ball's change in velocity is 22.7 m s^{-1} in the downward direction.

Q12 Magnitude of $v_2 - v_1 = \sqrt{12^2 + 10^2}$ = 15.6 m s^{-1}

$\theta = \tan^{-1}\left(\dfrac{10}{12}\right)$ = 39.8°

∴ Bearing = 230° to nearest degree

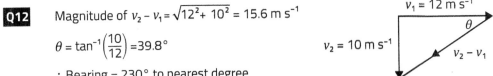

Q13 $\left[\dfrac{k}{m}\right] = \dfrac{\text{N m}^{-1}}{\text{kg}} = \dfrac{(\text{kg m s}^{-2})\text{m}^{-1}}{\text{kg}} = s^{-2}$, so $\left[\dfrac{m}{k}\right] = s^2$

Now, $[T]$ = s, so the first two of the given equations are clearly wrong.

And $[T^2]$ = s^2, so the third equation could be right, but the fourth is wrong.

Q14 (a) $A = \pi r^2 = \pi \times \left(\dfrac{0.317 \times 10^{-3} \text{ m}}{2}\right)^2$ = 7.89 × 10^{-8} m^2

(b) Volume = Al = 7.89 × 10^{-8} m^2 × 0.85 m = 6.71 × 10^{-8} m^3

(c) Volume = $\dfrac{\text{mass}}{\text{density}} = \dfrac{2.50 \text{ kg}}{8.96 \times 10^3 \text{kg m}^{-3}}$ = 2.790 × 10^{-4} m^3

$l = \dfrac{\text{volume}}{A} = \dfrac{2.790 \times 10^{-4} \text{m}^3}{7.893 \times 10^{-8} \text{m}^2}$ = 3.54 × 10^3 m = 3.54 km

Q15 (a) Pressure = $\dfrac{\text{(magnitude of) force normal to a surface}}{\text{area of surface}}$

So (magnitude of) force = pressure × area = 2.5 × 10^6 Pa × π (0.100 m)^2 = 7.9 × 10^4 N to 2 sf

(b) Upward force of gas on piston = weight of copper piston + downward force of air on piston.

So $pA = Al\rho g + p_A A$
in which A is the cross-sectional area of the copper piston, l is its length and ρ is the density of copper, so Al is the volume of the copper cylinder, and $Al\rho$ is its mass.

Dividing through by A, and then substituting numerical data

$p = l\rho g + p_A$
= 0.10 m × 8.96 × 10^3 kg m^{-3} × 9.81 m s^{-2} + 101 × 10^3 Pa
= 110 × 10^3 Pa

Q16 (a) Without the 4.17 cm width measurement, the range of width measurements is (4.28 – 4.24) cm = 0.04 cm; with the 4.17 cm it is 0.11 cm, so the 4.17 cm is well outside the range established by the other readings, and it is wise to discard it.

(b) The absolute uncertainty in the mass may be taken as 0.1 g, so its percentage uncertainty is (0.1 / 600) × 100 = 0.017 %. This is far lower than the percentage uncertainties in the lengths, widths and heights (see below) so can be ignored.

(c) Mean length, $l = \dfrac{6.35 + 6.38 + 6.34 + 6.38 + 6.37}{5} = 6.36$ cm

$\Delta l = \dfrac{6.38 - 6.34}{2}$ cm = 0.02 cm; so $p(l) = \dfrac{0.02}{6.36} \times 100\% = 0.31\%$

Mean width, $w = \dfrac{4.26 + 4.24 + 4.28 + 4.25}{4}$ cm = 4.26 cm

$\Delta w = \dfrac{4.28 - 4.24}{2}$ cm = 0.02 cm; so $p(w) = \dfrac{0.02}{4.26} \times 100\% = 0.47\%$

Mean height, $h = \dfrac{2.79 + 2.81 + 2.83 + 2.80 + 2.81}{5}$ cm = 2.81 cm

$\Delta h = \dfrac{2.83 - 2.79}{2}$ cm = 0.02 cm; so $p(h) = \dfrac{0.02}{2.81} \times 100\% = 0.71\%$

$\rho = \dfrac{m}{V} = \dfrac{599.5 \text{ g}}{6.36 \text{ cm} \times 4.26 \text{ cm} \times 2.81 \text{cm}} = 7.87$ g cm^{-3}

$p(\rho) = 0.31 + 0.47 + 0.71 = 1.5\%$

So $\Delta\rho = 7.87$ g cm^{-3} × ±0.015 = ±0.12 g cm^{-3},

i.e. $\rho = (7.87 \pm 0.12)$ g cm^{-3} or (7.9 ± 0.1) g cm^{-3}

[N.B. It's fine to work in kg and m, yielding $(7.87 \pm 0.12) \times 10^3$ kg m^{-3}.]

Q17 CoG of ruler is at 25.0 cm. If mass of ruler = m_R, using PoM (moments about the pencil):

0.100 kg × g × 0.140 m = $m_R g$ × 0.100 m

∴ m_R = 0.140 kg

Now, with the metal piece mass, m_m, moments about pencil again

$m_m g$ × 0.115 m = 0.140 kg × g × 0.125 m

∴ m_m = 0.152 kg

Q18 (a) The tensions in A and B are equal by symmetry: the links are the same distances from the edges of the sign, which is uniform. Each tension is equal to half the pull of gravity on the sign.

So tension = $\frac{1}{2}$ × 3.5 kg × 9.81 N kg^{-1} = 17.2 N

(b) A is 0.15 m from H, distance AB is 0.60 m, so B is 0.75 m from H. Centre of mass of bar is 0.45 m from H. Weight of bar = 1.5 kg × 9.81 N kg^{-1} = 14.7 N

Sum of clockwise moments = 17.2 N × 0.15 m + 17.2 N × 0.75 m + 14.7 N × 0.45 m
= 22.1 N m

(c) Sum of anticlockwise moments about H = sum of clockwise moments about H

So $T \cos (90° - 30°)$ × 0.75 m = 22.1 N m

So $T = \dfrac{22.1 \text{ N m}}{0.75 \text{ m } \cos 60°} = 59$ N

(d) F_{right}, the horizontal force component to the right of the hinge on the bar, must balance the horizontal component of the wire's force on the bar.

∴ $F_{right} = T \cos 30° = 59$ N × cos 30° = 51 N

If F_{up} is upward force component of hinge on bar, then

F_{up} = weight of bar + sum of tensions in A and B − upward component of T
= 14.7 N + 34.3 N − 59 N × cos (90° − 30°) = 19.5 N

∴ $F = \sqrt{51^2 + 19.5^2} = 55$ N at $\tan^{-1}\left(\dfrac{19.5}{51}\right) = 21°$ above the horizontal to the right

Q19 (a) Work is a scalar not a vector; displacement is a vector not a scalar.

(b) (i) $[\mu] = [\rho][u][L] = kg\,m^{-3} \times m\,s^{-1} \times m = kg\,m^{-1}\,s^{-1}$
$N\,s\,m^{-2} = (kg\,m\,s^{-2}) \times s \times m^{-2} = kg\,m^{-1}\,s^{-1}$
So the units of the right- and left-hand sides of the equation are the same.

(ii) $k = \dfrac{\rho u L}{\mu} = \dfrac{1.16\,kg\,m^{-3} \times 41.2\,m\,s^{-1} \times 0.071m}{1.87 \times 10^{-5}\,N\,s\,m^{-1}} = 1.8 \times 10^{5}$ (2 sf) / 1.81×10^{5} (3 sf)

Section 1.2: Kinematics

Q1 (a) (i) Mean speed = $\dfrac{\text{total distance travelled}}{\text{total time taken}}$

(ii) Mean velocity = $\dfrac{\text{total displacement}}{\text{total time taken}}$

(b) (i) Mean speed = $\dfrac{120\,m + 120\,m}{27\,s} = \dfrac{240\,m}{27\,s} = 8.9\,m\,s^{-1}$ (2 sf)

(ii) Total displacement = $120\sqrt{2}$ m on a bearing of 045° (in a NE direction)

∴ Mean velocity = $\dfrac{120\sqrt{2}}{27}\,m\,s^{-1} = 6.3\,m\,s^{-1}$ on a bearing of 045° (NE)

Q2 Taking east as the positive direction:
Initial velocity, $u = 19\,m\,s^{-1}$; final velocity $v = -11\,m\,s^{-1}$
$\Delta v = v - u = -11 - 19 = -30\,m\,s^{-1}$; time, $t = 25\,ms = 0.025\,s$

∴ Mean acceleration = $\dfrac{\Delta v}{t} = \dfrac{-30\,m\,s^{-1}}{0.025\,s}$ eastwards = $1200\,m\,s^{-2}$ westwards

This is a very large acceleration (about 120g) because the duration is so small.

Q3 (a) Using the labelled graph

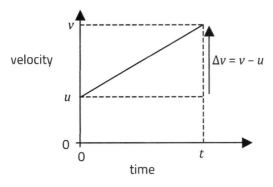

(i) Acceleration, $a = \dfrac{\Delta v}{t} = \dfrac{v - u}{t}$
Multiplying by t: $at = v - u$
Rearranging gives $v = u + at$ as required.

(ii) Displacement, x = area under the v–t graph = mean height × base
$= \dfrac{u + v}{2}t$

(iii) Displacement, x = area under the v–t graph
= area of rectangle + area of triangle
$= ut + \dfrac{1}{2}(v - u)t$
But $at = v - u$ so $x = ut + \dfrac{1}{2}at^2$

(b) Rearranging $v = u + at$ gives $t = \dfrac{v - u}{a}$.
Substitute for t in $x = \dfrac{u + v}{2}t \longrightarrow x = \dfrac{(u + v)(v - u)}{2a}t$
Multiply by $2a$ and expand the brackets $\longrightarrow 2ax = v^2 - u^2$. Hence $v^2 = u^2 + 2ax$.

Q4 (a) Taking upwards as positive

  (i) $u = 15.5$ m s^{-1}, $a = -g = -9.81$ m s^{-2}; $v = 0$ (at the top point); $x = h$ (max height)

$v^2 = u^2 + 2ax$, $\therefore 0 = (15.5)^2 - 2 \times 9.81h$

$\therefore$ max height $h = \dfrac{(15.5)^2}{2 \times 9.81} = 12.2$ m

Alternatively: use conservation of energy. $mgh = \frac{1}{2}mv^2$ and divide by m.

  (ii) $v = u + at$, so $t = \dfrac{v-u}{a} = \dfrac{0-15.5}{-9.81} = 1.58$ s

(b) The ball is decelerating, so the mean velocity in the first half of the ascent is greater than in the second. Hence the time taken to reach half the maximum height is less than half the time to reach the maximum.

(c) (i) Calculate time to drop 6.1 m from highest point.

Take downwards as positive: $u = 0$, $a = g = 9.81$ m s^{-2}, $x = 6.1$ m.

$x = ut + \frac{1}{2}at^2$, so $6.1 = \frac{1}{2} \times 9.81\, t^2$.

So $t = \sqrt{\dfrac{2 \times 6.1}{9.81}} = 1.12$ s.

$\therefore$ Total time = 1.58 s + 1.12 s = 2.70 s

Alternatively: calculate time from beginning with: $u = 15.5$ m s^{-1}, $a = -g$, $x = 6.1$ m
Use the same equation: $x = ut + \frac{1}{2}at^2$ which gives $6.1 = 15.5t - \frac{1}{2} \times 9.81\, t^2$
Solving the quadratic for $t \longrightarrow 0.46$ s and 2.70 s. Choose the second solution.

  (ii) Calculate the speed for the 6.1 m drop from the highest point:
Time to drop this far = 1.12 s, from (c)(i), taking downwards as positive
$v = u + at$, so $v = 0 + 9.81 \times 1.12 = 11.0$ m s^{-1}

Alternatively: Use $v = u + at$ from the beginning
With upwards positive $u = 15.5$ m s^{-1}, $a = -g$, $t = 2.70$ s
So $v = 15.5 - 9.81 \times 2.70 = -11.0$ m s^{-1} hence a downward velocity of 11.0 m s^{-1}

Or Calculate the speed gained in dropping 6.1 m using $v^2 = u^2 + 2ax$

Q5 (a) The water takes some time to reach the ground. When it is released it has a horizontal velocity. It keeps this horizontal motion when falling, so it needs to be released before getting to the drop zone.

(b) Let t = time to fall 100 m. Use $x = ut + \frac{1}{2}at^2$ with $u = 0$, $a = g$ and $x = 100$ m.

$\therefore t = \sqrt{\dfrac{2 \times 100}{9.81}} = 4.52$ s.

In 4.52 s, the plane travels 120 m s^{-1} $\times$ 4.52 s = 540 m (2 sf), so water must be dropped 540 m before the burning area.

Q6 (a) (i) $u_h = u \cos\theta = 20.0 \times \cos 37° = 16.0$ m s^{-1}

  (ii) $u_v = u \sin\theta = 20.0 \times \sin 37° = 12.0$ m s^{-1}

  (iii) The resultant of 16 m s^{-1} horizontally and 12 m s^{-1} vertically is not 16 + 12 = 28 m s^{-1} but $\sqrt{16^2 + 12^2} = 20$ m s^{-1}, because they are at right angles, so no worries!

(b) (i) Upwards is positive. $u_v = 12.0$ m s^{-1}, $a = -g$, $v_v = 0$, $x = h$ (height)
$v^2 = u^2 + 2ax$, $\therefore 0 = (12.0)^2 - 2 \times 9.81h$

$\therefore h = \dfrac{(12.0)^2}{2 \times 9.81} = 7.34$ m

  (ii) Let t = time to reach ground. Consider vertical motion
$x = ut + \frac{1}{2}at^2$, $\therefore 0 = 12.0t - 4.905t^2$, $\therefore t(12.0 - 4.905t) = 0$
$\therefore t = 0$ (ignore) or $4.905t = 12 \longrightarrow t = 2.45$ s
[Alternative method: twice the time to the maximum height.]
$\therefore$ Horizontal distance travelled = 16.0 m s^{-1} $\times$ 2.45 s = 39.2 m (3 sf)

(c) (i) The sin function is a ratio, so has no units.

$\therefore \left[\dfrac{u^2 \sin 2\theta}{g}\right] = \dfrac{(\text{m s}^{-1})^2}{\text{m s}^{-2}} = \text{m} = [R]$. So homogeneous.

(ii) Using the equation, $R = \dfrac{(20)^2 \sin 74°}{9.81} = 39.2$ m, i.e. the equation gives the same answer (at least to 3 sf).

Q7 (a) (i) From the graph, $v(20\ \text{s}) = 7.95\ \text{m s}^{-1}$; $v(10\ \text{s}) = 3.70\ \text{m s}^{-1}$

∴ Mean acceleration $= \dfrac{(7.95 - 3.70)\,\text{m s}^{-1}}{10.0\ \text{s}}\ 0.43\ \text{m s}^{-2}$ (to 2 sf)

(ii)

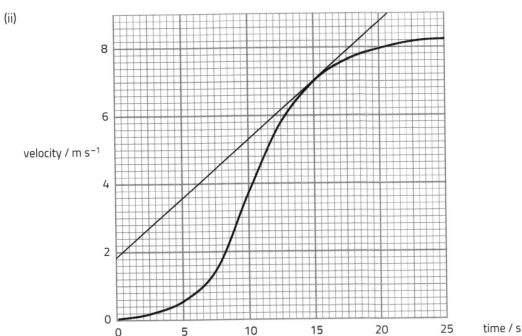

Acceleration at 15.0 s = gradient of tangent at 15.0 s

$$= \frac{(9.0 - 1.8)\,\text{m s}^{-1}}{(20.7 - 0.0)\,\text{s}}$$

$$= 0.35\ \text{m s}^{-2}.$$

(b) The chord to the graph from 14.5 s to 15.5 s is almost indistinguishable from the tangent at 15.0 s, so its gradient is almost identical. Hence to a good approximation this method should work. [However it is impossible to carry this out accurately because reading such small distances on the graph has a large fractional uncertainty.]

Q8 (a) Total displacement = area under graph

$$= \tfrac{1}{2} \times (15 + 29)\ \text{s} \times 14\ \text{m s}^{-1}$$

$$= 308\ \text{m}$$

∴ Mean velocity $= \dfrac{308\ \text{m}}{29\ \text{s}} = 10.6\ \text{m s}^{-1}$ (3 sf)

(b)

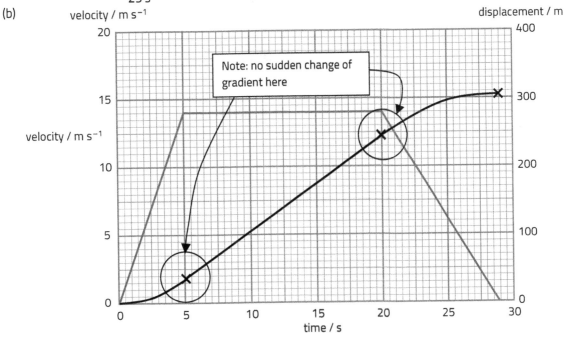

Calculations: Displacement at 5 s = 0.5 × 5 × 14 = 35 m
Displacement between 5 and 20 s = 15 × 14 = 210 m [⟶ total 245 m]
Displacement between 20 and 29 s = 0.5 × 9 × 14 = 63 m [⟶ total 308 m]

(c)

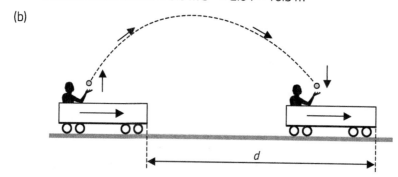

Q9 (a) Vertical motion; upwards positive.

Time ball in air $= \dfrac{v-u}{a} = \dfrac{(10.0 - (-10.0))\ \text{m s}^{-1}}{9.81\ \text{m s}^{-2}} = 2.04$ s.

∴ Distance train moves $= 8.0\ \text{m s}^{-1} \times 2.04 = 16.3$ m

(b)

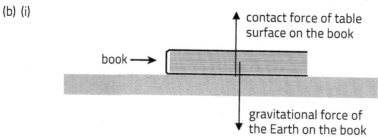

(c) When Helen throws the ball upwards, it also has a horizontal velocity. This horizontal velocity remains unchanged through the motion (ignoring air resistance) so when it comes down it has moved the same distance forward as Helen.

(d) From Helen's point of view, there is a backwards wind which pulls the ball behind her. From the observer's point of view, the ball is moving though the air, it experiences air resistance, which slows it down so it doesn't travel as far forward as Helen and lands behind her.

Section 1.3: Dynamics

Q1 (a) In an interaction between two bodies, A and B, the force exerted by body B on body A is equal in magnitude and opposite in direction to the force exerted by body A on body B.

(b) (i)

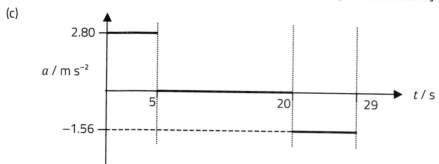

(ii) Contact force: N3 partner acts on the table surface
Gravitational force: N3 partner acts on the (whole) Earth

Q2 (a) Starting with Angharad's equation, the momentum, p, of a body is defined as mv (with the usual symbols). So for a constant mass:

$$F = \frac{\Delta p}{t} = \frac{\Delta(mv)}{t} = m\frac{\Delta v}{t}$$

Acceleration, a, is defined as $\dfrac{\Delta v}{t}$, so $F = ma$, which is Bethan's equation.

(b) There is not a single mass which is being accelerated, for which Bethan's equation would be useful. However, we can calculate the change in momentum of each molecule and therefore the total momentum change per second at the wall, so Angharad's is more useful.

Q3 (a) $v^2 = u^2 + 2ax \therefore a = \dfrac{v^2 - u^2}{2x} = \dfrac{(2.1 \text{ m s}^{-1})^2 - (1.5 \text{ m s}^{-1})^2}{2 \times 2.7 \text{ m}} = 0.40 \text{ m s}^{-2}$

Resultant force, $ma = 28 \text{ kg} \times 0.40 \text{ m s}^{-2} = 11.2 \text{ N}$

(b) If the frictional force on the box is F, the resultant force $= 18.2 \text{ N} - F$

$\therefore 18.2 \text{ N} - F = 11.2 \text{ N}$

$\therefore F = 18.2 \text{ N} - 11.2 \text{ N} = 7.0 \text{ N}$ to the left

So the frictional force of box on ground $= 7.0 \text{ N}$ to the right

Q4 (a) The resultant force on the trolleys together $= ma = 26 \text{ kg} \times 0.75 \text{ m s}^{-2} = 19.5 \text{ N}$

The total frictional force $= 10.0 \text{ N}$

$\therefore$ If $F =$ force exerted by rope, $F - 10.0 \text{ N} = 19.5 \text{ N}$

$\therefore F = 29.5 \text{ N}$

(b) Resultant force on trolley B $= 12 \text{ kg} \times 0.75 \text{ m s}^{-2} = 9.0 \text{ N}$

Frictional force $= 5.0 \text{ N}$

$\therefore$ force exerted by chain on trolley B $= 9.0 \text{ N} + 5.0 \text{ N} = 14.0 \text{ N}$ (forwards)

$\therefore$ By Newton's 3rd law, force exerted by B on chain $= 14.0 \text{ N}$ (backwards)

Alternatively: Resultant force on trolley **A** + chain $= 14 \text{ kg} \times 0.75 \text{ m s}^{-2} = 10.5 \text{ N}$

Frictional force $= 5.0 \text{ N}$ (backwards); force exerted by rope $= 29.5 \text{ N}$ (forwards)

$\therefore$ Backwards force exerted by **B** on chain $= 29.5 \text{ N} - 10.5 \text{ N} - 5.0 \text{ N} = 14.0 \text{ N}$

Q5 Magnitude of resultant force,

$R = \sqrt{8.0^2 + 16.0^2} = 17.9 \text{ N}$ (3 sf)

Direction: $\theta = \tan^{-1}\left(\dfrac{16.0}{8.0}\right) = 63.4°$

$\therefore$ Acceleration, $a = \dfrac{F}{m} = \dfrac{17.9 \text{ N}}{4.0 \text{ kg}} = 4.48 \text{ m s}^{-2}$

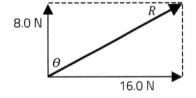

So the acceleration is 4.5 m s^{-2} on a bearing of $063°$
(both to 2 sf)

Q6 (a) Vertical components of force cancel.

Resultant horizontal force $= 2 \times 6.0 \cos 30° = 10.4 \text{ N}$ to the left.

$\therefore$ Acceleration $\dfrac{F}{m} = \dfrac{10.4 \text{ N}}{0.050 \text{ kg}} = 210 \text{ m s}^{-2}$ to the left (2 sf)

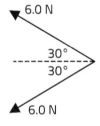

(b) Reason 1: the extension of the springs will decrease so the tension in them will also decrease.

Reason 2: The angle of the springs to the horizontal will increase so the horizontal component of the forces due to the tension will decrease.

Q7 (a) Let mass of each trolley $= m$.

Initial momentum to right $= 8.0m - 2 \times 2.2m = 3.6m$

$\therefore$ By conservation of momentum, momentum after collision $= 3.6m$ to the right

$\therefore$ Velocity of trolleys $= \dfrac{p}{\text{mass}} = \dfrac{3.6m}{3m} = 1.2 \text{ m s}^{-1}$ to the right.

(b) Initial kinetic energy $= \frac{1}{2}m \times (8.0)^2 + 2 \times \frac{1}{2}m \times (2.2)^2 = 36.84m$

Final kinetic energy $= 3 \times \frac{1}{2}m \times (1.2)^2 = 2.16m$

Fraction of initial kinetic energy $= \dfrac{2.16}{36.84} = 0.059$

(c) A very small fraction of the lost energy will be given out as sound waves. The collision will produce compression waves in the metal frames of the trolleys which will temporarily lead to increased vibrations of the metal ions (atoms), hence this is likely to be true. This additional energy will quickly be transferred to the air by conduction and convection.

Q8 The total momentum (strictly 'linear momentum') of the dumbbell is zero because momentum is a vector and the two weights are moving at equal speeds in opposite directions.
The kinetic energy of each weight = $\frac{1}{2}mv^2$ = 1.25 × (5.0)² = 31.25 J

Kinetic energy is a scalar and has no direction.
∴ Total kinetic energy = 31.25 + 31.25 = 63 J (2 sf)

Q9 (a) Considering energy transfer for the falling ball,
$\frac{1}{2}mv^2 = mgh$ ∴ Impact speed = $\sqrt{2gh} = \sqrt{2 \times 9.81 \times 2.00}$ = 6.26 m s^{-1}
Bounce speed = $\sqrt{2 \times 9.81 \times 1.20}$ = 4.85 m s^{-1}
Taking upwards as positive:
Impact velocity = –6.26 m s^{-1}; bounce velocity = 4.85 m s^{-1}
Change of velocity, Δv = 4.85 –(–6.25) = 11.1 m s^{-1} upwards
∴ Change of momentum, $\Delta p = m\Delta v$ = 0.220 ×11.1 = 2.44 N s = 2.4 N s (2 sf)

(b) Resultant force = $\frac{\Delta p}{t} = \frac{2.44 \text{ N s}}{0.150 \text{ s}}$ = 16.3 N

(c) There are two vertical forces on the ball: the weight, W (= 0.220 × 9.81 = 2.2 N), which acts downwards, and the contact force, C, by the ground, which acts upwards.
∴ Resultant force $R = C - W$, so $C = R + W = R + 2.2$ N.

Q10 (a) Take motion to the right as positive. Applying the principle of conservation of momentum, sum of mv terms is constant:
$$0.15 \times 0.36 - 0.25 \times 0.52 = 0.15v + 0.25 \times 0.107$$
where v = velocity of 0.15 kg glider after the collision.
∴ 0.15 v = –0.10275
∴ v = –0.685 m s^{-1}, i.e. 0.685 m s^{-1} (3 sf) to the left.

(b) Initial kinetic energy = $\frac{1}{2}$ × 0.15 × (0.36)² + $\frac{1}{2}$ × 0.25 × (0.52)² = 0.0435 J
Final kinetic energy = $\frac{1}{2}$ × 0.15 × (–0.685)² + $\frac{1}{2}$ × 0.25 × (0.107)² = 0.0366 J
∴ There is a loss of kinetic energy, hence inelastic.

Alternatively: Closing speed (before collision) = 0.36 + 0.52 = 0.88 m s^{-1}
Separating speed (after collision) = 0.685 + 0.107 = 0.792 m s^{-1}
0.792 m s^{-1} < 0.88 m s^{-1}, so inelastic.

Q11 (a) Normal component of velocity changes from –2500 cos 60° to +2500 cos 60°.
So Δv = 2500 m s^{-1}.
So $\Delta p = m \Delta v$ = 6.6 × 10^{-27} kg × 2500 m s^{-1} = 1.65 × 10^{-23} N s (upwards)

(b) To get back to side XY, molecule must travel vertically by 6.0 cm + 6.0 cm = 12.0 cm. The vertical component of velocity = ±1250 m s^{-1}.
∴ Time between collisions on XY = $\frac{0.120 \text{ m}}{1250 \text{ m s}^{-1}}$ = 9.6 × 10^{-5} s
∴ Mean force on XY = $\frac{1.65 \times 10^{-23} \text{ N s}}{9.6 \times 10^{-5} \text{ s}}$ = 1.7 × 10^{-19} N

Section 1.4: Energy concepts

Q1 (a) Power = energy per unit time = $\frac{\text{energy}}{\text{time}}$; so [energy] = [power] × [time]
The hour (h) is a unit of time, so [energy] = kW h

(b) 96 kW h = 96 × 10³ W × 3 600 s
= 3.5 × 10⁸ J (2 sf)

Q2 (a) Energy can neither be created nor destroyed but can be transferred from one form (or body) to another.

(b) (i) Initial kinetic energy = $\frac{1}{2}mv^2$ = 0.5 × 0.150 × (50.0)² = 187.5 J

Gain of gravitational potential energy = mgh = 0.150 × 9.81 × 31.9 = 46.9 J

∴ Kinetic energy at 31.9 m = 187.5 − 46.9 = 140.6 J

∴ 0.5 × 0.150 × v^2 = 140.6

∴ Speed, v = 43 m s⁻¹ (2 sf)

(ii) No, all the values of energy contain the factor m = 0.150 kg, so it could be changed or cancelled out without affecting the calculation.

Q3 (a) Loss in GPE = 85 kg × 9.81 N kg⁻¹ × 200 m sin 5°

= 14 500 J

(b) Gain in kinetic energy = $\frac{1}{2}mv^2$ = 0.5 × 85 kg × (12.0 m s⁻¹)² = 6 120 J

∴ Loss of mechanical energy = 14 500 J − 6 100 J = 8 400 J

∴ Work done against friction = 8 400 J, so frictional force × 200 m = 8 400 J,

∴ Frictional force = $\frac{8\,400\ \text{J}}{200\ \text{m}}$ = 42 N (2 sf)

Q4 (a) Loss of kinetic energy = work done against friction

∴ F × 55 m = 0.5 × 1 200 kg × (26.7 m s⁻¹)²

= 428 000 J

∴ F = 7 800 N (2 sf)

(b) $F_{30} = \dfrac{0.5 \times 1200 \times (13.35)^2}{14\ \text{m}}$ = 7 600 N (2 sf); $F_{50} = \dfrac{0.5 \times 1200 \times (22.25)^2}{38\ \text{m}}$ = 7 800 N (2 sf)

$F_{70} = \dfrac{0.5 \times 1200 \times (31.15)^2}{75\ \text{m}}$ = 7 800 N (2 sf) (Velocities converted with factor 60 mph = 26.7 m s⁻¹)

All the forces are very close, so the assumption is true.

[Note: An alternative, and more insightful, method is to compare the ratios v^2/d and show that they are all very close. There is no need to convert mph to m s⁻¹.]

(c) John is incorrect. The kinetic energy loss is proportional to the mass of the car, so the braking force must also be proportional to the mass.

Q5 (a) The energy possessed by a system is defined as the amount of work it can do. When a system does 10 J of work, then 10 J of energy is transferred. Hence the two definitions are the same.

(b) In the case of the horse, the energy transfer is effected by a force moving its point of application. Hence the work definition is useful. In the case of the Sun, the energy transfer is from individual emissions of photons, a situation in which force is not a useful concept.

Q6 (a) Initial GPE = mgh

= 0.600 kg × 9.81 N kg⁻¹ × 0.400 m = 2.35 J.

(b) Loss in GPE = 0.600 × 9.81 × 0.184 = 1.08 J

Elastic potential energy, EPE = $\frac{1}{2}kx^2$ = 0.5 × 32.0 N m⁻¹ × (0.184 m)² = 0.54 J

Kinetic energy = 1.08 J − 0.54 J = 0.54 J

(c) $\frac{1}{2}$ × 0.600 kg × v^2 = 0.54 J

∴ $v = \sqrt{\dfrac{2 \times 0.54\ \text{J}}{0.600\ \text{kg}}}$ = 1.34 m s⁻¹ (3 sf)

(d) At the lowest point, KE = 0, so: loss of GPE = gain of EPE

If the distance fallen = x, $mgx = \frac{1}{2}kx^2$, ∴ $x = \dfrac{2mg}{k} = \dfrac{2 \times 0.600\ \text{kg} \times 9.81\ \text{N kg}^{-1}}{32\ \text{N m}^{-1}}$

∴ x = 0.368 m

∴ Height above bench = 0.400 − 0.368 = 0.032 m

(e) (i) mgh = loss of GPE; $\frac{1}{2}kh^2$ = gain in EPE; $\frac{1}{2}mv^2$ = gain in KE (because initial KE = 0)

The equation is an expression of energy conservation: the loss of GPE is equal to the gain in EPE + the gain in KE

(ii) Inserting values: $0.6 \times 9.81h = 16h^2 + 0.3$

$\therefore 16h^2 - 5.886h + 0.3 = 0$

$\therefore h = \dfrac{5.886 \pm \sqrt{5.886^2 - 4 \times 16 \times 0.3}}{2 \times 16} = 0.061$ m or 0.306 m

Q7 (a) From the definition of velocity, $x = vt$, where t = time.
From the definition of power, $W = Pt$.
Substituting for W and x in $W = Fx \longrightarrow Pt = Fvt$ and hence $P = Fv$.

(b) (i) $[k] = $ N $($m s$^{-1})^{-2} = $ kg m s^{-2} (m^{-2} s^2)

$= $ kg m^{-1}

(ii) Velocity is constant, so drag D = forward force F
$P = Fv = (kv^2)\,v$

$= 0.4$ kg m^{-1} $\times$ (30 m s$^{-1})^3$

$= 10\,800$ W

(iii) [Note: in this question, your answer will depend very much on your assumptions]
Assume a steady speed of 30 m s^{-1} and an efficiency of 80%

Time taken for 900 km $= \dfrac{900 \times 10^3}{30}$ s $= 30\,000$ s

$\therefore$ Energy transfer $= 10\,800$ W $\times 30\,000$ s $= 3.2 \times 10^8$ J
80% of 100 kW h $= 0.8 \times 100 \times 3.6 \times 10^6$ J $= 2.9 \times 10^8$ J.
This suggests that the claim is incorrect but not very far off – perhaps the assumed driving speed was less.

Q8 (a) Ignoring the angle to the horizontal, $W = 83$ N $\times 7.0 \times 10^3$ m

$= 5.8 \times 10^5$ J (2 sf)

(b) Strictly, $W = Fx \cos \theta$, where θ = angle to the horizontal.
But, if $\theta < 5°$, $\cos \theta > 0.9962$, so for an answer to 2 sf (like the data) the actual angle is irrelevant.

(c) 83 N

(d) The sledge is moving at constant velocity, so the resultant force on the sledge is zero, so there is no energy transfer to the sledge. Energy is transferred to the ice as thermal energy.

(e) (i) Kinetic energy $\longrightarrow$ thermal energy (internal energy)

(ii) Kinetic energy of sledge $= \frac{1}{2} \times 210$ kg $\times$ (1.4 m s$^{-1})^2 = 205.8$ J

Distance travelled $= \dfrac{\text{work done}}{\text{frictional force}} = \dfrac{\text{drop in kinetic energy}}{\text{frictional force}} = \dfrac{205.8 \text{ J}}{83 \text{ N}} = 2.5$ m (2 sf)

(f) The resultant force $= 105$ N $- 83$ N $= 22$ N, so nearly 4 times as much of the work is done in overcoming friction as in accelerating the sledge.

Q9 (a) (i) $[\text{LHS}] = $ kg s^{-1}
$[\text{RHS}] = $ m^2 (m s^{-1}) kg m$^{-3} = $ kg s$^{-1} = [\text{LHS}]$, so homogeneous

(ii) The power input is the KE input per second $= \frac{1}{2}mv^2$

$= \frac{1}{2}\,\pi r^2 v \rho \times v^2$

$= \frac{1}{2}\,\pi r^2 \rho v^3$

(iii) $P_{\text{OUT}} = 0.56 \times 0.95 \times 0.85 \times \frac{1}{2}\pi \times$ (6.0 m)$^2 \times 1.3$ kg m$^{-3} \times$ (15 m s$^{-1})^3$

$= 110\,000$ W (2 sf)

$= 110$ kW

(b) The mean power output $\propto v^3$ so at half the wind speed, the power output would be one eighth of the value in (a)(iii), so this would not be enough. But to calculate the mean power output we need to know the mean cube speed, e.g. if on two days, the speeds were 0 and 15 m s^{-1}, the mean power output (for the two days) would be 55 kW, which would still be too small and Bethan is probably right.

Section 1.5: Solids under stress

Q1 (a) F = tension (or force); k = spring constant; x = extension

(b) $k = \dfrac{F}{x}$, so $[k] = \dfrac{\textbf{kg m s}^{-2}}{\textbf{m}} = $ kg s^{-2}

(c) The extension is within the elastic limit.

Q2 Crystalline: one in which the atoms are arranged in a regular array, with long-range order, e.g. diamond.
Amorphous: one in which the arrangement of the atoms has no long-range order, e.g. glass.
Polymeric: one with long chain molecules consisting of multiple repeat units, e.g. polythene.

Q3 (a)

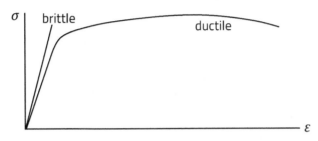

(b) [One answer] Concrete can be made stronger under tension by pre-compressing it. Metal rods under tension are placed in the concrete. When the tension is released, the concrete is placed under compression. Hence any cracks cannot propagate.

Q4 (a) Using $mg = kx$, $k = \dfrac{mg}{x} = \dfrac{0.300 \text{ kg} \times 9.81 \text{ N kg}^{-1}}{0.151 \text{ m}} = 19.49$ N m^{-1}

Fractional uncertainties: $p_m = \dfrac{6}{300} = 0.020$; $p_x = \dfrac{0.2}{15.1} = 0.013$

$\therefore p_k = 0.020 + 0.013 = 0.033$

Absolute uncertainty: $\Delta k = 0.033 \times 19.49 = 0.6$ (1 sf)

$\therefore k = (19.5 \pm 0.6)$ N m^{-1}

(b) Method: 1. Pull the mass (300 g) down a few cm and release it.

2. Use a stop watch to measure the time for (say) 20 oscillations.

3. Repeat 2 several times, calculate the mean value and estimate the uncertainty.

4. Calculate the period and the uncertainty by dividing by the number of oscillations.

5. Calculate T and its uncertainty by putting m and k in the equation and comparing with the measured value of the period..

Q5 (a) A ductile material is one that can be drawn into a wire (or can be deformed plastically without becoming brittle).

(b) (i) Consisting of a large number of interlocking crystals.

(ii) The diagram shows planes of atoms in a crystal. There is an extra half plane (ending at Y). This kind of defect is called an edge dislocation. When the tensile force, F, is applied, the bonds at **X** break and new ones are formed with **Y**. The dislocation moves to the right and causes a change in shape that does not reverse when the tension is removed.

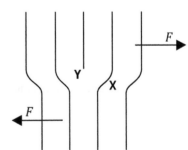

Q6 Elastic strain is a deformation which is reversed when the applied force is removed.
Stress is the tension divided by the cross-sectional area of the material.

The elastic limit is the value of stress below which the material exhibits elastic strain only and above which it exhibits plastic strain.

Plastic strain is strain which is not reversed when the applied force is removed.

Q7

$\dfrac{F_A}{F_B} = 1$	$\dfrac{\varepsilon_A}{\varepsilon_B} = 6$	$\dfrac{W_A}{W_B} = 3$
$\dfrac{\sigma_A}{\sigma_B} = 4$	$\dfrac{\Delta l_A}{\Delta l_B} = 3$	$\dfrac{W_A/V_A}{W_B/V_B} = 24$

Calculation

The tension is the same throughout, so $F_A = F_B$. Area of B = 4 × area of A, so $\sigma_A = 4\sigma_B$.

$$\varepsilon_A = \frac{\sigma_A}{E_A} = \frac{4\sigma_B}{E_B/1.5} = 6\varepsilon_B \quad \therefore \frac{\varepsilon_A}{\varepsilon_B} = 6; \quad \Delta l_A = \varepsilon_A l_A = 6\varepsilon_B \times \tfrac{1}{2}l_B = 3\Delta l_B .$$

$$W_A = F_A \Delta l_A = F_B \times 3\Delta l_B = 3W_B; \quad \frac{W_A}{V_A} = \frac{3W_B}{\frac{1}{8}V_B} = 24\frac{W_B}{V_B}$$

Q8 (a) To remove kinks from the wire

(b) $E = \dfrac{\sigma}{\varepsilon} = \dfrac{Fl_0}{A\Delta l} = \dfrac{4Fl_0}{\pi D^2 \Delta l}$

$\Delta D = \dfrac{D_{max} - D_{min}}{2} = \dfrac{0.25 - 0.23}{2} = 0.01$; so $\therefore$ % uncertainty in $D = \dfrac{0.01}{0.24} \times 100\% = 4.2\%$

$\therefore p(A) = 2 \times 4.2\% = 8.4\%$; $p(\Delta l) = \dfrac{0.1}{3.1} \times 100\% = 3.2\%$

The uncertainties in length and the force are negligible, so the percentage uncertainty in
$E = 8.4\% + 3.2\% = 11.6\% = 10\%$ (1 sf) or 12% (2 sf)

(c) If the D is (say) halved, A is divided by 4 and so σ is multiplied by 4, so Δl is multiplied by 4. $p(A)$ is × 2 and $p(\Delta l)$ is ÷ 4. In this case $p(E)$ would be 16.8% + 0.8% ~ 18% which is a greater uncertainty.

Q9 (a) The stress at which large plastic strain occurs (or at which edge dislocations move).

(b) $\sigma = 60$ MPa $\therefore \varepsilon = \dfrac{\sigma}{E} = \dfrac{60 \text{ MPa}}{2.00 \text{ GPa}} = 3.0 \times 10^{-4}$

EPE $= \tfrac{1}{2} Fx = \tfrac{1}{2} \times 60 \times 10^6$ Pa $\times \pi (0.015 \text{ m})^2 \times 3.0 \times 10^{-4} \times 5\,000$ m

$= 3.2 \times 10^4$ J

Q10 (a) Work done $= \tfrac{1}{2} F\Delta x = \tfrac{1}{2} \times 280$ N $\times 0.76$ m $= 106.4$ J $= 110$ J (2 sf)

(b) KE given to arrow = 90% × 106.4 J = 95.76 J

$\therefore$ velocity $= \sqrt{\dfrac{2E_k}{m}} = \sqrt{\dfrac{2 \times 95.76 \text{ J}}{0.050 \text{ kg}}} = 61.9$ m s^{-1}

Vertical component (= horizontal component) of velocity = 61.9 cos 45° = 43.8 m s^{-1}

Time in air $= \dfrac{v - u}{a} = \dfrac{43.8 - (-43.8)}{9.81} = 8.92$ s.

$\therefore$ Range (using horizontal velocity) = 43.8 m s^{-1} × 8.92 s = 390 m (2 sf), so the claim is slightly exaggerated (unless there is some aerodynamic lift).

Q11 All the kinetic energy of the train needs to be transferred to the elastic energy in the springs before the springs reach maximum compression, i.e. $E_k = \tfrac{1}{2} Fx = \tfrac{1}{2} kx^2$. The smaller the spring constant, the larger the distance needed to stop the train. A large distance needs a long housing for the spring which is a disadvantage.

However, $F = \sqrt{\dfrac{2E_k}{x}}$, so this longer decelerating distance results in a lower maximum force, so a smaller deceleration of the train, i.e. a less violent halt, which is an advantage.

Section 1.6: Using radiation to investigate stars

Q1 Power $= \dfrac{\text{energy transfer}}{\text{time}}$, $\therefore$ W = J s^{-1}.

$\therefore$ W m^{-2} = (kg m^2 s^{-2}) s^{-1} m^{-2}

$= $ kg s^{-3}

Q2 A black body is one which absorbs all electromagnetic radiation which is incident upon it. No body emits more radiation [due to its temperature] at any wavelength than a black body.

Q3 $L = \sigma A T^4 = \sigma(4\pi r^2)T^4$,

$$\therefore \frac{L_{\text{red dwarf}}}{L_{\text{Sun}}} = \left(\frac{r_{\text{red dwarf}}}{r_{\text{Sun}}}\right)^2 \left(\frac{T_{\text{red dwarf}}}{T_{\text{Sun}}}\right)^4$$

$$\therefore L_{\text{red dwarf}} = \left(\frac{1}{4}\right)^2\left(\frac{1}{2}\right)^4 \times 4 \times 10^{26} \text{ W} = \frac{4 \times 10^{26} \text{ W}}{256} = 1.6 \times 10^{24} \text{ W}$$

Q4 (a) $L = \left(\frac{0.7 M_\odot}{M_\odot}\right)^4 L_\odot = 0.24\, L_\odot$

(b) For a number $x < 1$, the value of x^n decreases as n increases. So, in fact, n should be greater than 4, and Alex is incorrect.

[Note: $(0.7)^5 = 0.17$, which is closer to the actual value.]

Q5 Assuming that the star emits as a black body, the wavelength, λ_{max}, of the peak spectral intensity relates to the kelvin temperature, T, by Wien's displacement law: $\lambda_{\text{max}} = \frac{W}{T}$, where W is a constant, 2.9×10^{-3} m K. Hence the temperature of the star's photosphere (its surface) can be determined. The intensity, I, of the radiation from the star can be determined from the total area under the graph. The luminosity, L, of the star can then be calculated using $L = 4\pi d^2 \times I$, where d is the distance to the star.

The luminosity, L is given by the Stefan–Boltzmann law: $L = A\sigma T^4$, where A is the surface area of the star. Hence, A can be calculated (knowing T from Wien's law) and thus the radius, r, from $A = 4\pi r^2$.

Q6 The continuous emission spectrum is the radiation which is given out at all wavelengths. The line absorption spectrum is the missing radiation at particular wavelengths in the spectrum.

Q7 Atoms in the star's atmosphere can be promoted to higher states by the absorption of radiation with a photon energy equal to the difference in energy levels. This results in the dark lines (Fraunhofer lines) which are characteristic of the elements of the atoms in the star.

Q8 (a) $T = \frac{W}{\lambda_{\text{max}}} = \frac{2.90 \times 10^{-3}\text{ m K}}{501 \text{ nm}} = 5790$ K

(b) If the Sun were a black body of temperature 5790 K, its luminosity would be:

$$L = 4\pi\left(\frac{d}{2}\right)^2 \sigma T^4 = 4\pi \times \left(\frac{1.39 \times 10^9 \text{ m}}{2}\right)^2 \times (5.67 \times 10^{-8}\text{ W m}^{-2}\text{ K}^{-4}) \times (5790\,\text{K})^4$$

$$= 3.87 \times 10^{26} \text{ W}$$

This is very close to the website value so it is consistent with the Sun emitting as a black body.

Q9 (a) (i) $\lambda_{\text{max}} = \frac{W}{T} = \frac{2.90 \times 10^{-3}\text{ m K}}{4\,000\text{ K}} = 725$ nm

(ii) Near infra-red (very close to the red end of the visible spectrum).

(b) The ratio of the e-m power emitted per unit area from the sunspot compared to that of the Sun's photosphere is given by $\left(\frac{4000}{6000}\right)^4 = 0.20$ (2 sf). Also, nearly all the radiation is emitted in the infra-red – the Sun emits mainly in the visible – so any visible image will contain very little radiation from the sunspot.

Q10 Photon energy $E_{\text{ph}} = hf = \frac{hc}{\lambda}$

Wien's law: $\lambda_{\text{max}} = \frac{W}{T}$, photon energy $E_{\text{ph max}} = \frac{hc}{\lambda_{\text{max}}} = \frac{hc}{W}T$. So Bryn is correct.

Q11 Multiwavelength astronomy is taking images of objects in the universe using several regions of the electromagnetic spectrum, e.g. X-rays, visible, radio. The different regions give information about different conditions and processes that occur, e.g. comparing UV, visible and infra-red images of stars enables us to compare their temperatures.

Unit 1 Answers

Q12 $\lambda_{max} = \frac{W}{T}$. For cosmic background radiation, $\lambda_{max} \sim 1$ mm and galactic molecular clouds have $\lambda_{max} \sim 0.1$ mm, so using microwave and submillimetre radiation gives us information about processes there.

λ_{max} for stars like the Sun (6000 K) $\sim$ 500 nm, so the range for Red Giant to Blue Giant stars is ~ 1 µm – 50 nm, which spans the IR and UV regions of the e-m spectrum.
Supernovae, black holes and inter-galactic gas have $\lambda_{max} \sim 10^{-9} - 10^{-11}$ m, which is in the X-ray region of the spectrum, so X-ray astronomy can reveal processes in these objects.

Q13 $E_{ph} = hf = \frac{hc}{\lambda}$

For 10 nm $E_{ph} = \dfrac{6.63 \times 10^{-34} \times 3.00 \times 10^8}{10 \times 10^{-9}} = 2.0 \times 10^{-17}$ J $= \dfrac{2.0 \times 10^{-17} \text{J}}{1.60 \times 10^{-19} \text{J/eV}} = 120$ eV

For 400 nm, the values are 1/40 of these, i.e. 5.0×10^{-19} J, 3.0 eV

Q14 (a) Photon energy $= \dfrac{6.63 \times 10^{-34} \times 3.00 \times 10^8}{1.0 \times 10^{-6} \times 1.60 \times 10^{-19}} = 1.2$ eV . If there are He$^+$ ions in the 3rd excited state (energy –3.4 eV) they can absorb photons of energy 1.2 eV and enter the 4th excited state (energy –2.2 eV) because the difference in energy levels is 1.2 eV. The light in the direction of the emission from the star is depleted in photons of this energy, giving rise to the dark line.

(b) Transitions from –6.0 eV $\longrightarrow$ –3.4 eV require 2.6 eV and –3.4 $\longrightarrow$ –1.5 require 1.9 eV, both of which are in the visible range. (Note: Ionisation from the –2.2 eV level will also absorb visible photons but this will not be a single line because there is no single upper energy level.)

(c) In order to produce these lines, the He$^+$ ions must first be excited to the –6.0 eV energy level, which requires an energy of 48.4 eV. Photons in the visible range have energies around 2–3 eV and the Sun's spectrum (from its 6 000 K surface) only extends a small way into the UV, so there is not enough energy to do this and Eleri is correct.

Section 1.7: Particles and nuclear structure

Q1 Electrons have no structure – they cannot be separated into constituent particles.
Protons are composed of 3 quarks: 2 up quarks and 1 down quark.

Q2 The particle is a hadron: it is composed of quarks and/or anti-quarks

Q3 Strong: proton, pi+ meson, anti-neutron
Electromagnetic: electron, proton, pi+ meson, anti-neutron, positron
Weak: all of them

Q4 (a) Electromagnetic
(b) (i) The decay time is appropriate for e-m interaction (intermediate)
   (ii) Photons are produced (so cannot be strong)

Q5 (a) (i) positron, e$^+$
   (ii) anti-neutron, $\bar{n}$
   (iii) anti-electron neutrino (or electron anti-neutrino), $\bar{\nu}_e$
(b) The π^- has quark structure $d\bar{u}$; the π^+ has structure $u\bar{d}$. So the quarks in π^+ are the anti-quarks to those in π^-, so Eurig is correct.

Q6 (a) It cannot be a lepton because there are no leptons with $Q = 2$. A hadron with $Q = 2$ must have 3 quarks because no combination of a quark and anti-quark, which all have charges $\pm\frac{1}{3}$ or $\pm\frac{2}{3}$, can have $Q = 2$. So it must be a baryon.
(b) (i) Strong interaction because of the short time-scale
   (ii) Baryon number, charge, quark flavour, lepton number

(iii) $\Delta^{++} \longrightarrow p + \pi^+$. At the quark level, this is uuu $\longrightarrow$ uud + u$\bar{\text{d}}$
Charge: $Q(\Delta^{++}) = 2$; $Q(p) + Q(\pi^+) = 1 + 1 = 2$, $\therefore$ conserved.
Baryon number: $B(\Delta^{++}) = 1$; $B(p) + B(\pi^+) = 1 + 0 = 1$, $\therefore$ conserved.
Quark flavours: $U(\Delta^{++}) = 3$; $U(p) + U(\pi^+) = 2 + 1 = 3$ $\therefore$ conserved.
$D(\Delta^{++}) = 0$; $D(p) + D(\pi^+) = 1 + (-1) = 0$ $\therefore$ conserved
Lepton number: no leptons involved, so $L = 0$ on left and right, $\therefore$ conserved

Q7 (a) π^+ is a meson; e^+ is a lepton; v_e is a lepton

(b) Charge is conserved (the pion and positron are both +; the neutrino is neutral)
Lepton number is conserved: $L(\pi^+) = 0$; $L(e^+) + L(v_e) = -1 + 1 = 0$
Baryon number is conserved: 0 for all particles

(c) Quark flavour is not conserved: For the π^+, $U = 1$, $D = -1$. For the positron and neutrino both U and D are zero.

Q8 (a) Charge: the left-hand side has $Q = 1 + 1 = 2$; right-hand side $Q = = 1 + (-1) + 1 = 1$
Baryon number: $B(p + p) = 1 + 1 = 2$; $B(\Delta^+ + e^- + \pi^+) = 1 + 0 + 0 = 1$
Up number: $U(p + p) = 2 + 2 = 4$; $B(\Delta^+ + e^- + \pi^+) = 2 + 0 + 1 = 3$
Down number: $D(p + p) = 1 + 1 = 2$; $D(\Delta^+ + e^- + \pi^+) = 1 + 0 + (-1) = 0$
Lepton number: $L(p + p) = 0 + 0 = 0$; $L(\Delta^+ + e^- + \pi^+) = 0 + 1 + 0 = 1$

(b) This decay, producing a photon, is an electromagnetic decay, which is slower than the first decay, which is a strong decay.

Q9 Lepton number $L(n + \pi^+) = 0$; $L(\Delta^{++} + e^-) = 0 + 1 = 1$
Charge: $Q(n + \pi^+) = 0 + 1 = 1$; $Q(\Delta^{++} + e^{-1}) = 2 - 1 = 1$. So not violated.
Baryon number: $B(n + \pi^+) = 1 + 0 = 1$; $Q(\Delta^{++} + e^-) = 1 - 0 = 1$. So not violated.

Q10 (a) Baryons have a baryon number of 1. 3 quarks have a baryon number of $\frac{1}{3} + \frac{1}{3} + \frac{1}{3} = 1$, e.g. the neutron has quark structure udd.

(b) Mesons each have a quark, with baryon number $\frac{1}{3}$, and an antiquark with baron number $-\frac{1}{3}$, e.g. the pion, π^+, has quark structure u$\bar{\text{d}}$. So the total baryon number is $\frac{1}{3} - \frac{1}{3}$, i.e. 0.

Q11 The protons and electrons can interact via the electromagnetic interaction, which has a long range. They can lose energy to electrons in the atoms through which they pass. The protons can also interact with nucleons in the lead nuclei via the strong force, but this has a minor effect as the range of the force is only about 10^{-15} m.
The neutrinos can interact with electrons and nucleons only with the weak force. Hence they have to approach within $\sim 10^{-17}$ m and the probability of an interaction is very small.

Q12 **Lepton number.** Leptons are fundamental particles. The lepton family consists of 3 generations. The first generation members are the electron and the [electron] neutrino. They each have a lepton number, L, of 1. The antiparticles to these (the positron and the anti-neutrino) have lepton numbers of -1. In any interaction, the total lepton number is conserved, e.g. if an electron is the only reacting particle, either an electron or a neutrino must be a product particle.

Baryon number. Baryons comprise 3 quarks, which are fundamental particles, and each has a baryon number, B, of 1. Anti-baryons, comprising 3 anti-quarks, have $B = -1$. In any interaction the baryon number is conserved.

Charge: This is always conserved in particle interactions.

Quark flavour: First-generation quarks have two flavours, up (u) and down (d), each with a separate flavour number, U and D. U and D are conserved in strong and electromagnetic interactions but may change by ± 1 in weak interactions.

Q13 (a) The decay of a $\Delta^+ \longrightarrow p + \pi^0$ conserves charge ($Q = 1$), baryon number ($B = 1$) and lepton number ($L = 0$) and the mass of the Δ^+ is greater than p + π^0 together.
The mass of p + ρ^0 ($3352m_e$) is greater than the mass of the Δ^+.

(b) In the decay $\pi^0 \longrightarrow e^- + e^+ + \gamma$, the charge, baryon number and lepton number are all conserved – they are all 0 on both sides of the equation. Hence the decay is possible (by the electromagnetic interaction). The mass of the product particles ($2m_e$) is less than the mass of the pion.

(c) Baryon number must be conserved. The only baryon less massive than the neutron is the proton, so the neutron must decay into a proton.

The reaction $n \longrightarrow p + \pi^-$ is impossible because the total mass of $p + \pi^-$ ($2043m_e$) is greater than that of the neutron.

So the neutron can only decay into a proton and an electron ($m = 1837m_e$). In order to conserve lepton number, an electron anti-neutrino must also be produced. Hence it must be a weak decay.

(d) To preserve baryon number, it would have to decay into another baryon but the proton is the lightest baryon, so this is impossible.

Practice questions: Unit 2: Electricity and Light

Section 2.1: Conduction of electricity

Q1 $Q = It = 0.015\ \text{A} \times 60\ \text{s} = 0.90\ \text{C}$

No. of electrons $= \dfrac{\text{charge}}{e} = \dfrac{0.90\ \text{C}}{1.60 \times 10^{-19}\ \text{C}} = 5.6 \times 10^{18}$ (2 sf)

Q2 [An electrical conductor is] a material which allows charge to flow through it.

Q3 Re-arranging the equation: $C = \dfrac{Q^2}{2W}$; the number 2 has no unit.

$\therefore F = [C] = \dfrac{[Q^2]}{[W]} = \dfrac{[Q]^2}{[W]} = \dfrac{A^2\ s^2}{\text{kg m}^2\ \text{s}^{-2}} = \text{kg}^{-1}\ \text{m}^{-2}\ \text{s}^4\ \text{A}^2$

Q4 Charge on an α-particle $= 2e = 3.20 \times 10^{-19}\ \text{C}$

$\therefore$ Current $= 3.20 \times 10^{-19}\ \text{C} \times 37 \times 10^3\ \text{s}^{-1} = 1.2 \times 10^{-14}\ \text{A}$

Q5 $v = \dfrac{I}{nAe}\ ,\ \therefore\ \dfrac{v_A}{v_B} = \dfrac{I_A}{I_B} \times \dfrac{n_B}{n_A} \times \dfrac{A_B}{A_A} = \dfrac{I_A}{I_B} \times \dfrac{n_B}{n_A} \times \left(\dfrac{d_B}{d_A}\right)^2$

$\dfrac{v_A}{v_B} = \dfrac{1.5}{10} \times \dfrac{1.0 \times 10^{29}}{3.0 \times 10^{29}} \times \left(\dfrac{0.30}{0.60}\right)^2 = 0.013$ (2 sf)

Q6 (a) The intensity of the light at the LDR is inversely proportional to the distance^2, so the number of photons reaching the LDR per second $\propto$ distance^{-2}. Assuming the number of conduction electrons is proportional to the number of photons per second, Nigel is correct.

(b)

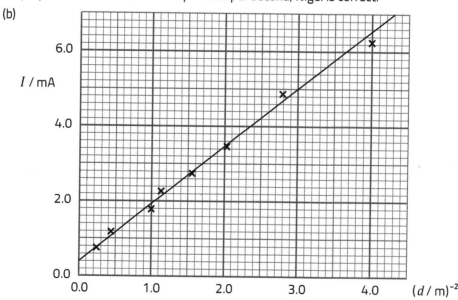

The graph of I against d^{-2} is expected to be a straight line through the origin. The points lie on quite a good straight line of positive gradient with a low degree of scatter, which agrees with Nigel's suggestion. However, the line does not pass through the origin. A possible reason for that is that the experiment was carried out in a room with some background light.

Section 2.2: Resistance

Q1 Current = rate of charge flow: $= I = \dfrac{Q}{t}, \therefore Q = It.$

pd = energy per unit charge flow $= \dfrac{W}{Q} = \dfrac{W}{It} = \dfrac{300 \text{ J}}{1.5 \text{ A} \times 20 \text{ s}} = 10 \text{ V}$

Q2 Current, $I = \dfrac{Q}{t}, \therefore [Q] = [I][t] = \text{A s}$

pd = energy per unit charge flow, so $V = \text{J C}^{-1} = (\text{kg m}^2 \text{ s}^{-2})(\text{A s})^{-1} = \text{kg m}^2 \text{ s}^{-3} \text{ A}^{-1}$

Q3 (a) For a metallic conductor at a constant temperature the current is directly proportional to the potential difference.

(b) The equation $V = IR$ is a statement of Ohm's law only if it is also stated that the resistance, R, is a constant. Also the conditions should be stated (metallic conductor and constant temperature).

Q4 (a) n = number of free electrons per unit volume; A = cross-sectional area of conductor
e = electronic charge; v = free-electron drift velocity

(b) If the temperature of the wire increases, so does the mean kinetic energy of the free electrons. Their random speeds are therefore greater and so the time between collisions with the lattice ions is less. Because of this the increase in velocity due to the pd is less than at lower temperatures and so is the drift velocity, making the current less. A lower current for the same pd means that the resistance is greater.

Q5 (a) Superconductivity is the property of zero electrical resistance, which occurs when the conductor is below a certain temperature – the transition temperature , T_c.

(b) A high-temperature superconductor is one with a transition temperature above the boiling point of liquid nitrogen – about $-200\,°\text{C}$.

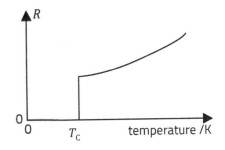

Advantage: it costs much less to achieve a superconducting state, using liquid nitrogen than conventional superconductors, which require liquid helium for cooling, especially in high-field industrial magnets such as in MRI.

Q6 (a) (i) $R_{2.4V} = \dfrac{2.4 \text{ V}}{1.5 \text{ A}} = 1.6\ \Omega;\qquad R_{12V} = \dfrac{12 \text{ V}}{3.0 \text{ A}} = 4.0\ \Omega$

$\therefore \dfrac{\text{Resistance of the filament at 12 V}}{\text{Resistance of the filament at 2.4 V}} = 2.5$

(ii) $\dfrac{P_{12V}}{P_{2.4V}} = \dfrac{12 \text{ V} \times 3.0 \text{ A}}{2.4 \text{ V} \times 1.5 \text{ A}} = 10$

(b) As the pd increases, the free electrons have a greater acceleration between collisions, so gain more kinetic energy. Hence more energy is transferred to the lattice ions when they collide.

Q7 (a) Consider 1.00 m of wire.

$\rho = \dfrac{RA}{I} = \dfrac{2.18\ \Omega}{1.00 \text{ m}} \times \pi \left(\dfrac{0.1 \times 10^{-3} \text{ m}}{2} \right)^2 = 1.7 \times 10^{-8}\ \Omega \text{ m}$

(b) Volume of 1 kg (14 306 m length) of wire $= \pi \left(\dfrac{0.1 \times 10^{-3} \text{ m}}{2} \right)^2 \times 14\,306 \text{ m} = 1.12 \times 10^{-4} \text{ m}^3$

$\therefore$ Density $= \dfrac{\text{mass}}{\text{volume}} = \dfrac{1.00 \text{ kg}}{1.12 \times 10^{-4} \text{ m}^3} = 8900 \text{ kg m}^{-3} = 8.9 \text{ g cm}^{-3}$, so in good agreement.

Q8 (a) $\rho = \dfrac{RA}{l} = \dfrac{13.9\,\Omega}{2.000\,\text{m}} \times \pi \left(\dfrac{0.32 \times 10^{-3}\,\text{m}}{2}\right)^2 = 5.589 \times 10^{-7}\,\Omega\,\text{m}$

Fractional uncertainties: $p_d = \dfrac{0.01}{0.32} = 0.03125,\ \therefore p_A = 2 \times 0.03125 = 0.0625$

$p_l = \dfrac{0.002}{2.0} = 0.001;\ p_R = \dfrac{0.1}{13.9} = 0.007 \therefore p_\rho = 0.0625 + 0.001 + 0.007 = 0.0705$

$\therefore$ Absolute uncertainty: $\Delta_\rho = 0.0705 \times 5.589 \times 10^{-7} = 0.4 \times 10^{-7}$

$\therefore \rho = (5.6 \pm 0.4) \times 10^{-7}\,\Omega\,\text{m}$

(b) For double the diameter the p_d is half and p_A is one quarter of the new value, so ~0.031. The resistance would also be about one quarter, so with the same uncertainty in R, p_R would be 4 times the original, that is 0.028 and so the total p would be about 0.06, which is slightly, but only slightly, less. The uncertainty in length hardly affects the total uncertainty, but the fractional uncertainty in resistance will be halved (from 0.007 to 0.0035) which has a bigger effect.

Q9 $P = \dfrac{V^2}{R}$, so, when operating, $R = \dfrac{V^2}{P} = \dfrac{(240\,\text{V})^2}{60\,\text{W}} = 960\,\Omega.$

Taking room temperature to be 290 K: operating temperature $= \dfrac{960\,\Omega}{80\,\Omega} \times 290\,\text{K} = 3500\,\text{K}$ (2 sf)

Q10 (a) Because the resistance is constant, the current taken from the supply and therefore the power of the heater will also be constant.

(b) $P = \dfrac{V^2}{R}$, so $R = \dfrac{V^2}{P} = \dfrac{(30\,\text{V})^2}{10\,\text{W}} = 90\,\Omega$

$R = \dfrac{\rho l}{A}$, so $l = \dfrac{RA}{\rho} = \dfrac{90\,\Omega}{4.9 \times 10^{-7}\,\Omega\,\text{m}} \times \pi \left(\dfrac{0.12 \times 10^{-3}\,\text{m}}{2}\right)^2 = 2.1\,\text{m}$ (2 sf)

Q11 $R_A = \dfrac{4\rho l}{\pi D^2}$

$R_B = \dfrac{4(2.5\rho)(3l)}{\pi(2D)^2} = \dfrac{7.5}{4} \times \dfrac{4\rho l}{\pi D^2} = 1.875\,R_A$

$P = \dfrac{V^2}{R}$. The voltages are the same, so $\dfrac{P_A}{P_B} = \dfrac{R_B}{R_A} = 1.875$

Q12 (a) The wire coil is connected to a multimeter on its resistance range. The test tube is placed in a beaker with a water / ice mixture and allowed to equilibrate. A thermometer is placed in the beaker and the resistance and temperature is measured. The beaker is gradually heated using a bunsen burner and the resistance measured at a range of temperatures up to 100°C.

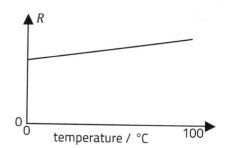

(b) Assuming that the resistance varies linearly with temperature:

Increase in resistance per degree $= \dfrac{16.3\,\Omega - 12.0\,\Omega}{75\,°\text{C} - 20\,°\text{C}} = 0.0782\,\Omega\,°\text{C}^{-1}$

$\therefore$ Temperature of oil $= 20\,°\text{C} + \dfrac{18.7\,\Omega - 12.0\,\Omega}{0.0782\,\Omega\,\text{K}^{-1}} = 85.7\,\text{K}$

so temperature of oil $= 20\,°\text{C} + 85.7\,°\text{C} = 106\,°\text{C}$

Q13

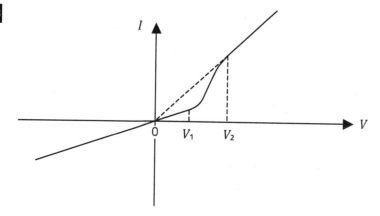

Section 2.3: DC circuits

Q1 (a) With the switch open, the lamps L_1 and L_2 are in series. Hence, the current is the same in the two lamps, so they have the same brightness.

(b) When S is closed, L_2 and L_3 are in parallel, so the resistance of this combination is less than that of L_2 on its own. Hence the resistance of the circuit is less and so the current through L_1 is greater and it is brighter. The pd across L_1 is also greater so the pd across L_2 is less, so the current is less and so L_2 is dimmer. L_3 has the same brightness as L_2 because they are identical and have the same current.

Q2 (a) For each coulomb of charge which flows through the two resistors, 5 J of energy is transferred in the left-hand resistor and 3 J in the right. So the total energy transfer per coulomb is 8 J. This must also be the energy transferred in the battery so the pd is 8 V.

(b) 8 V.

Q3 $R = 12\,\Omega \longrightarrow I = 0.50\ \text{A}$ $R = 24\,\Omega \longrightarrow I = 0.25\ \text{A}$

$R = 36\,\Omega \longrightarrow I = 0.17\ \text{A}$

$R = 6.0\,\Omega \longrightarrow I = 1.0\ \text{A}$ $R = 4.0\,\Omega \longrightarrow I = 1.50\ \text{A}$

$R = 18\,\Omega \longrightarrow I = 0.33\ \text{A}$

$R = 8.0\,\Omega \longrightarrow I = 0.75\ \text{A}$

Q4 (a) Current in $12\,\Omega$ resistor $= 0.30\ \text{A} \times \dfrac{20\,\Omega}{12\,\Omega} = 0.50\ \text{A}$

∴ Current in $15\,\Omega$ resistor $= 0.30\ \text{A} + 0.50\ \text{A} = 0.80\ \text{A}$

$P = I^2 R$,

so total power $= (0.80)^2 \times 15 + (0.30)^2 \times 20 + (0.50)^2 \times 12 = 14.4\,\Omega$

(b) The resistance of the $15\,\Omega$ / X combination is less than that of the $15\,\Omega$ alone, so the total resistance of the circuit decreases, so the total current increases. Hence the pd across the $20\,\Omega$ / $12\,\Omega$ combination increases and so the pd across the $15\,\Omega$ decreases.

Q5 (a) The top component in the circuit is an LDR. As the light level increases, the resistance of the LDR decreases. Hence, the total resistance of the circuit decreases and so the current increases. The pd across the fixed resistor, which is equal to V_{OUT}, increases.

(b) At 37°C, resistance of thermistor $= 6.4\ \text{k}\Omega$.

The thermistor needs to be in the bottom position because its resistance increases as the temperature decreases, thus increasing the output voltage.

Using potential divider to calculate the resistance, R, of the series resistor:

$5\ \text{V} = \dfrac{6.4\ \text{k}\Omega}{R + 6.4\ \text{k}\Omega} \times 12\ \text{V}$ ∴ $R = 9.0\ \text{k}\Omega$ (2 sf)

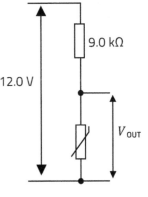

Q6 (a) 9.0 J of energy is transferred from chemical store per coulomb of charge passing through the battery.

(b) (i)   Chemical energy transferred per second $= 9.0\ \text{V} \times 1.5\ \text{A} = 13.5\ \text{W}$.

Energy transferred by the electric current in the external circuit $= 7.8\ \text{V} \times 1.5\ \text{A} = 11.7\ \text{W}$

Energy transferred to thermal energy by the electric current inside the battery is the difference in these, i.e. 13.5 W – 11.7 W = 1.8 W

(ii) $V = E - Ir$ So $r = \dfrac{E - V}{I} = \dfrac{9.0\ V - 7.8\ V}{1.5\ A} = 0.8\ \Omega$ [or, use $P_r = I^2 r$, with P_r = 1.8 W and I = 1.5 A]

Q7

With one 10 Ω resistor, the pd is 6.5 V, so $I = \dfrac{6.5\ V}{10\ \Omega} = 0.65$ A

Applying $V = E - Ir \longrightarrow 6.5 = E - 0.65r$ (1)

With two 10 Ω resistors, the total external resistance is 5 Ω, and the pd is 6.0 V

so $I = \dfrac{6.0\ V}{5.0\ \Omega} = 1.2$ A $\longrightarrow 6.0 = E - 1.2r$ (2)

Subtracting equation (2) from equation (1) $\longrightarrow 0.5 = 0.55r$

$\therefore r = 0.91\ \Omega$

Substituting in (1) and rearranging $\longrightarrow E = 6.5 + 0.65 \times 0.91 = 7.1$ V

Q8

$V = E \times \dfrac{R}{R + r}$ and $I = \dfrac{E}{R + r}$; when $R = r$, $V = \dfrac{E}{2}$ and $I = \dfrac{E}{2r}$, so $P_{out} = \dfrac{E^2}{4r}$

For the conventional cell: $P_{out} = \dfrac{(1.5)^2}{4 \times 0.3} = 1.875$ W

For the Ni-Cd cell, $P_{out} = \dfrac{(1.2)^2}{4 \times 0.035} = 10.3$ W = 5.5 × the max power from the conventional cell.

Q9 (a) 18 Ω resistor in parallel with a series combination of 3.3 Ω and 10 Ω.

$R = \dfrac{18 \times (3.3 + 10)}{18 + (3.3 + 10)}\Omega = \dfrac{239.4}{31.3}\Omega = 7.6\ \Omega$ (2 sf)

(b) Eliminating V from the equations: $E - Ir = IR$

$\therefore E = I(R + r)$

Dividing by $EI \longrightarrow \dfrac{1}{I} = \dfrac{R}{E} + \dfrac{r}{E}$

So a graph of $1/I$ against R is a straight line of gradient $1/E$ and intercept on the $1/I$ axis of r/E.

(c) 2.3   2.9   3.2   3.7   4.3   5.3

(d)

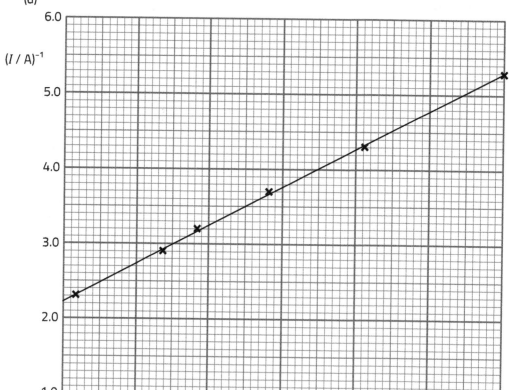

(e) Gradient = $\dfrac{5.3 - 2.2}{18.0 - 3.0} = 0.207\ V^{-1}$, giving a value of E of $\dfrac{1}{0.207\ V^{-1}} = 4.84$ V

Applying $y = mx + c$, then $c = y - mx$. Using the values (10.5, 3.76) gives $c = 1.59$ A^{-1}.

Then $r = \dfrac{\text{intercept}}{\text{gradient}} = \dfrac{1.59 \text{ A}^{-1}}{0.207 \text{ V}^{-1}} = 7.7 \ \Omega$.

This shows that the teacher's value of emf is consistent but the internal resistance is higher than the student's results.

Q10 (a) pd across the terminals of the battery $= 3 \times 1.2$ V $= 3.6$ V

Current $= \dfrac{1.5 \text{ W}}{2.5 \text{ V}} = 0.6$ A. pd across resistor $= 3.6$ V $- 2.5$ V $= 1.1$ V

$\therefore$ Resistance of resistor, $R = \dfrac{1.1 \text{ V}}{0.6 \text{ A}} = 1.8 \ \Omega$

(b) Fraction $= \dfrac{I^2 R}{I^2 R + I^2 R_{\text{lamp}}} = \dfrac{R}{R + R_{\text{lamp}}} = \dfrac{1.8 \ \Omega}{1.8 \ \Omega + (2.5/0.6) \ \Omega} = 0.30$

Q11 (a) $V_{\text{total}} = 0.70$ V $+ 0.03$ A $\times 820 \ \Omega = 25.3$ V

(b) Resistance needed for 10 mA $= \dfrac{(9.0 - 1.9)\text{V}}{0.010 \text{ A}} = 790 \ \Omega$

Resistance needed for 25 mA $= \dfrac{(9.0 - 1.9)\text{V}}{0.025 \text{ A}} = 316 \ \Omega$

So 470 Ω and 680 Ω are suitable

Section 2.4: The nature of waves

Q1 Waves involve no transfer of material; the particles in the medium just oscillate about a fixed point.

Q2 In a transverse wave, the particles of the medium oscillate at right angles to the direction of propagation. In a longitudinal wave, the particles of the medium oscillate parallel to the direction of propagation.
Examples: Transverse – seismic S waves; Longitudinal – seismic P waves.

Q3 (a) In a polarised light beam all the oscillations (of the electric field) are in the same direction.
(b) If a beam is partially polarised, oscillations occur in all directions at right angles to the direction of propagation but a fraction of the waves oscillate in the same direction.

Q4 (a) In an unpolarised beam, all directions have the same energy, so the energy in any component is 50% of the total.
(b) The unpolarised part has an intensity of 0.4 W m^{-2}, giving a constant intensity 0.2 W m^{-2}. The polarised part has an intensity of 0.6 W m^{-2}, so the transmitted intensity of this part should oscillate between 0 and 0.6 W m^{-2} with the maximum and minimum 90° apart. So the total intensity should oscillate between 0.2 and 0.8 W m^{-2}, 90° apart, which agrees with the graph and Cheryl is correct.

Q5

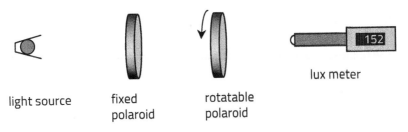

light source / fixed polaroid / rotatable polaroid / lux meter

The apparatus is set up as shown in the diagram. The rotatable polaroid is rotated through a set of positions, e.g. 15° apart. At each position the reading on the lux meter is noted and a graph plotted of reading against angle. The reading on the meter varies sinusoidally between a maximum and a minimum value. The angle between the two maxima is 180°. The graph is expected to look as follows.

Expected results:

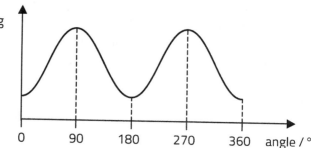

light-meter reading

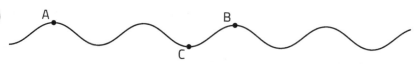

Q6 (a) (i)

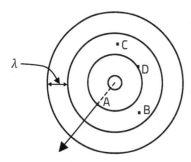

(ii) NB. In part (i), any peak; in part (ii), any trough.

(b) (i) $4\lambda = 3.04$ m, so $\lambda = 0.76$ m

(ii) Amplitude = 3.7 mm

(iii) $6T = 0.46$ s, so period = 0.0767 s

$$\therefore \text{speed} = \frac{0.76\,\text{m}}{0.0767\,\text{s}} = 9.9 \text{ m s}^{-1}$$

Q7 (a)

(b) Wavelength = $\frac{18.0\,\text{cm}}{3} = 6.0$ cm; frequency = $\frac{25}{8.0\,\text{s}} = 3.125$ Hz

Speed = $\lambda f = 6.0 \times 3.125 = 19$ cm s^{-1} (2 sf)

Q8 a) P waves: $\lambda = \frac{v}{f} = \frac{6.2\,\text{km s}^{-1}}{8.9\,\text{Hz}} = 0.70$ km [= 700 m] (2 sf)

S waves: $\lambda = \frac{v}{f} = \frac{3.7\,\text{km s}^{-1}}{8.9\,\text{Hz}} = 0.42$ km [= 420 m] (2 sf)

(b) Let distance to Bala = d (in km)

Time for S waves from Liberty Stadium to Bala = $\frac{d}{3.7}$; time for P waves = $\frac{d}{6.2}$

$\therefore$ Time delay = $\frac{d}{3.7} - \frac{d}{6.2} = d\left(\frac{1}{3.7} - \frac{1}{6.2}\right) = 0.109d$

$\therefore 0.109d = 16.3, \therefore d = \frac{16.3}{0.109} = 150$ km.

Q9 (a) The direction of travel is at right angles to the wavefronts. The direction of oscillation is at right angles to both the direction of travel and the wavefronts.

[Note: These ripples are actually 'surface waves' in which the motion of a particle is a vertical circle, with the plane of the circle in the direction of propagation. However, you will lose no marks if you treat them as transverse.]

(b) Number of wavelengths = 7.0, so $\lambda = \frac{14.6\,\text{cm}}{7} = 2.086$ cm

$f = \frac{20}{4.7\,\text{s}} = 4.255$ Hz

$\therefore$ Speed = $\lambda f = 2.086$ cm × 4.255 Hz = 8.9 cm s^{-1}

(c) The number of waves passing any point in the shallow water must be the same as in the deep water (they aren't spontaneously created or destroyed), i.e. the frequency stays the same. The wavelength is given by $v = \lambda f$, so as the speed decreases, so does the wavelength and Gerallt is correct in both these statements.

Section 2.5: Wave properties

Q1   Diffraction is the spreading out of waves after they pass through a gap or past the edge of an obstacle.

Q2   When the slit is 300 nm wide, the waves spread out by 90° on both sides of the slit but the intensity of the transmitted light is very low. As the width increases (towards 600 nm) the intensity increases – more in the forward direction. Above 600 nm the transmitted waves become concentrated into an increasingly narrow central band with weaker side bands separated by narrow channels without any waves.

Q3   (a) The waves are produced by two dippers which vibrate in phase at the centres of the circles. The two sets of waves pass through each other. At points where the waves are in phase (e.g. at B), they add to give larger amplitude waves. At points where the waves are out of phase (e.g. at A and C) they subtract and cancel each other out.

   (b) (i) 2λ

   (ii) 1.5λ [or $1\frac{1}{2}\lambda$, if you prefer]

   (c) There is a constant phase difference between the oscillations of the sources.

Q4   (a) The total displacement of two waves which pass through the same point is the [vector] sum of the displacements of the individual waves.

   (b) (i)

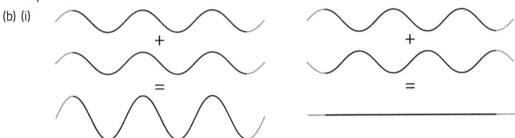

The left diagram shows constructive interference. The waves add to give one of double the amplitude. The right-hand diagram shows destructive interference. The waves add to give zero amplitude.

   (ii)

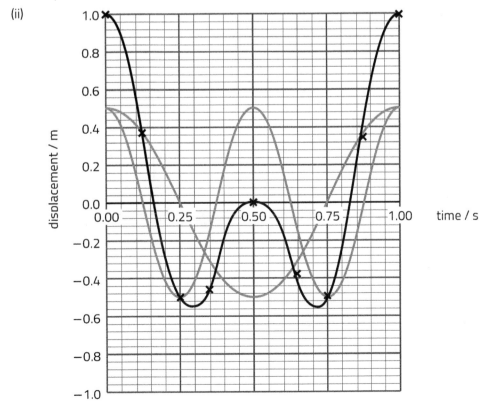

Q5 (a) It established that light was a wave phenomenon (not particles, as Newton had supposed).

(b)

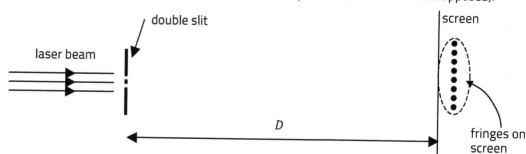

A blackened slide with two parallel slits about 0.5 mm apart is prepared and the separation, x, of the slits is measured using a travelling microscope. The slits are illuminated with a laser beam, as shown, and the fringes viewed on a distant screen ($D > 1$ m). The distance, D, to the screen is measured using metre rules or a tape measure. The separation, Δy, of the fringes is determined by measuring the distance between the outermost fringes using a mm scale and dividing by the number of fringe gaps visible. Δy is determined for a range of distances D, and a graph plotted of Δy against D.

The relationship between Δy and D is $\Delta y = \dfrac{\lambda D}{x}$, so the graph should be a straight line through the origin. The gradient $m = \dfrac{\lambda}{x}$, so the gradient, m, is determined and λ calculated using $\lambda = mx$.

Q6 (a) The microwaves incident on S_1 and S_2 spread out by diffraction and the divergent beams overlap at P and interfere. The waves are in phase at S_1 and S_2 and the distance S_1P and S_2P are equal, so the waves are in phase at P. Hence they interfere constructively, producing a maximum.

(b) For destructive interference, $S_2Q - S_1Q = \left(n + \frac{1}{2}\right)\lambda$ where λ is the wavelength of the microwaves and $n = 0, 1, 2\ldots$ As it is the first such point above P, $n = 0$ so $S_2Q - S_1Q = \frac{1}{2}\lambda$

For constructive interference, $S_2R - S_1R = n\lambda$. For R, $n = 1$ so, $S_2R - S_1R = \lambda$.

(c) (i) $\Delta y = \dfrac{\lambda D}{a}$, so PQ $= \dfrac{\lambda D}{2a} = \dfrac{2.8 \text{ cm} \times 15.0 \text{ cm}}{2 \times 6.5 \text{ cm}} = 3.23$ cm

(ii) Using Pythagoras, with PR = 6.5 cm:

$$\frac{\lambda}{2} = S_2Q - S_1Q = \sqrt{15.0^2 + (3.23 + 3.25)^2} - \sqrt{15.0^2 + (3.23 - 3.25)^2} = 1.34 \text{ cm}$$

This gives a value of 2.7 cm for the wavelength, which is 4% too small – but still quite close!

Q7 (a)

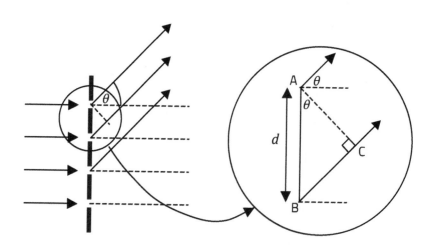

For the light from the top two slits to arrive at a distant point in phase, the path difference must be $n\lambda$ ($n = 0, 1, 2, \ldots$). On the expanded diagram, the angle BAC $= \theta$.

∴ Path difference = BC = $d \sin\theta$, where d = distance between slits.

So $n\lambda = d \sin\theta$.

Note: if this is true for two neighbouring slits, it is true for all slits.

(b) (i) When $n = 1$, $\theta = \tan^{-1}\left(\dfrac{0.225\text{ m}}{1.750\text{ m}}\right)$, $\therefore \sin\theta = 0.1275$

$$d = \frac{1}{250 \times 10^3 \text{ m}^{-1}} = 4.00 \times 10^{-6}\text{ m}$$

$\therefore \lambda = \dfrac{d\sin\theta}{n} = 4.00 \times 10^{-6} \times 0.1275 = 5.10 \times 10^{-7}\text{ m}$

(ii) Maximum possible value of $\sin\theta = 1.0$ $\therefore n_{max} = \dfrac{d}{\lambda} = \dfrac{4.00 \times 10^{-6}\text{ m}}{5.10 \times 10^{-7}\text{ m}} = 7.8$

n_{max} must be an integer, so possible values $= 0, \pm1, \pm2, \ldots\ldots\pm7$

$\therefore$ 15 bright dots.

Q8 (a) [see diagram right]

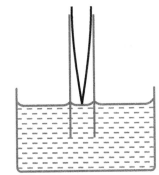

(b) Length $l = \dfrac{\lambda}{4}$, $\therefore \lambda = 4l$.

$c = \lambda f = 4lf$

(c) 317 m s^{-1}

309 m s^{-1}

All the values of c are lower than the true value ranging from 7% less at 256 Hz to 12% at 480 Hz. The % inaccuracy has a steady increasing trend as the frequency increases, suggesting a systematic uncertainty.

(d) $l = \dfrac{c}{4f}$, so 4 × gradient of a graph of l against $\dfrac{1}{f}$ should be c.

No grid available, so taking the extreme values: when $l = 0.312$ m, $\dfrac{1}{f} = 3.91 \times 10^{-3}$ s

When $l = 0.157$ m, $\dfrac{1}{f} = 2.08 \times 10^{-3}$ s

$\therefore$ 4 × gradient $= 4 \times \dfrac{(0.312 - 0.157)\text{ m}}{(3.91 - 2.08) \times 10^{-3}\text{ s}} = 339$ m s^{-1}, which is within 1% of the true value.

Section 2.6: Refraction of light

Q1 (a) At the boundary between two given materials, the ratio of the sine of the angle of incidence to the sine of the angle of refraction is a constant.

(b) For light passing from a vacuum into the material, $n = \dfrac{\sin i}{\sin r}$ where i is the angle of incidence and r is the angle of refraction.

(c) $n = \dfrac{c}{v}$, where v is the speed of light in the material and c the speed of light in a vacuum.

Q2 $n = \dfrac{c}{v}$, so $v = \dfrac{c}{n} = \dfrac{3.00 \times 10^8}{1.49}$ m s$^{-1} = 2.01 \times 10^8$ m s^{-1}

Q3 (a) Because electrons cannot go faster than the speed of light in a vacuum.

(b) (i) Speed of light in water $= \dfrac{c}{n} = \dfrac{3.00 \times 10^8}{1.33} = 2.26 \times 10^8$ m s^{-1}. Any faster than this and Cherenkov radiation will be produced.

(ii) Using $\frac{1}{2}mv^2 = eV$, $V = \dfrac{mv^2}{2e} = \dfrac{9.11 \times 10^{-31} \times (2.26 \times 10^8)^2}{2 \times 1.60 \times 10^{-19}} = 145$ kV

Q4 (a) $v_{air} > v_{water} > v_{glass}$. Hence the angle, θ, to the normal in the three materials follows the same pattern: $\theta_{air} > \theta_{water} > \theta_{glass}$.

(b) $1.00 \sin 45° = 1.52 \sin x$ $\therefore$ $x = \sin^{-1}\left(\dfrac{\sin 45°}{1.52}\right) = \sin^{-1} 0.4652 = 27.7°$

$y = x$ [alternate angles] $= 27.7°$.

$1.52 \sin 27.7° = 1.33 \sin z$, $\therefore$ $z = \sin^{-1}\left(\dfrac{1.52 \sin 27.7°}{1.33}\right) = \sin^{-1} 32.1°$

(c) $n\sin\theta$ is constant, so $1.00 \sin 45° = 1.33 \sin z$ and the glass is irrelevant (apart from holding the water in place).

Q5 (a) $r = \sin^{-1}\left(\dfrac{\sin 27°}{1.42}\right) = 18.6°$

(b) Angle of incidence on back surface = 18.6° (angles in a segment) so angle of refraction back into air = 27°.

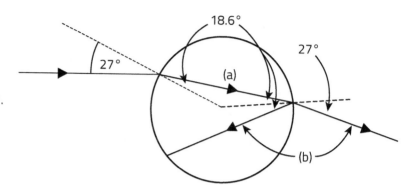

Q6 (a) Light travelling in a medium is incident on another medium, in which the speed of light is higher, with an angle of incidence greater than the critical angle.

(b) Critical angle, $c = \sin^{-1}\left(\dfrac{1}{n}\right) = \sin^{-1}\left(\dfrac{1}{1.55}\right) = 40.2°$.

(c) Angle of refraction = 19.8° (by considering angles in the triangle)
∴ $i = \sin^{-1}(1.55 \sin 19.8°) = 31.7°$

(d) The angle of incidence on the bottom surface is 19.8° (see diagram), which is less than the critical angle so total internal reflection does not occur and some of the light will emerge and Briony is incorrect. In fact the light emerges at the same angle as it entered the prism (31.7°).

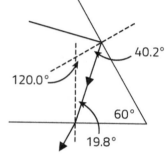

Q7

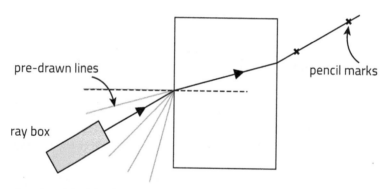

Place the glass block in the middle of a large piece of paper and draw its outline. Mark a series of lines at a regular set of angles to the normal (e.g. 15°, 30°, 45°, 60°, 75°) as shown and shine a narrow beam from a ray box along one of them. Put marks at well-separated points on the emergent ray as shown. Repeat for each of the other pre-drawn line. Remove the block and use the outline, the pencil marks and a ruler to reconstruct each of the paths for the light through the block. For each line, use a protractor to measure the angles of incidence, i, and refraction, r, at the point of incidence.

Plot sin i against sin r and draw a best-fit line, which should be straight and through the origin. Determine the gradient of the line – this is the refractive index.

Q8 When the level of the benzene is as shown, the ray of light passes straight through the prism into the benzene, almost undeviated, because the refractive indices of the benzene and glass are almost the same. If the level of benzene drops below the 'minimum height' mark the light ray is totally reflected on both lower surfaces (first horizontal and then vertical) and emerges back out and is detected by the alarm. This occurs because the angle of incidence (45°) is greater than the critical angle for the glass:
$c = \sin^{-1}\left(\dfrac{1}{1.5}\right) = 42°$.

Q9 (a) $\theta = \sin^{-1}\left(\dfrac{1.00 \times \sin 21.2°}{1.63}\right) = 12.8°$

$\phi = 90° - \theta = 77.2°$.

(b) The critical angle for the boundary between the fibre core and the cladding is given by:

$$c = \sin^{-1}\left(\frac{1.60}{1.63}\right) = 79.0°$$

The angle of incidence is less than the critical angle so total internal reflection does not occur.

(c) (i) Distance travelled = $\frac{14.0\,\text{km}}{\cos 5°}$ = 14.053 km. So the extra distance = 53 m.

(ii) The extra distance of 53 m takes light a time of $\frac{53\,\text{m}}{3.00 \times 10^8\,\text{m s}^{-1}}$ = 1.8 × 10⁻⁷ s.

The signal consists of a series of pulses of light. So for pulse intervals greater than about 10^{-7} s (i.e. a pulse frequency of 10 MHz) the pulses which travel by different paths will overlap with earlier or later pulses and the pulses will be unreadable – this is multi-mode dispersion.

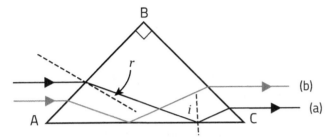

(a) The angle of refraction, $r = \sin^{-1}\left(\frac{1.00 \times \sin 45°}{1.60}\right)$ = 26.2°. The angle of incidence, i on AC is 71.2°, which is greater than the critical angle and hence TIR occurs. The angle of incidence on BC is 26.2° so the angle of refraction is 45°, i.e. the emergent ray is parallel to the initial ray.

(b) (See grey rays on diagram.) The angles are the same so the light rays are parallel but the upper one on the left is the lower one on the right.

(c) If the refractive index were, say, 1.3, the angle of incidence on AC would be 77° so, yes there would be TIR if the light ray hits AC. However, the lower the refractive index, the larger the angle of refraction on AB and if it is large enough, the ray wouldn't hit AC at all. So James is right, but for low refractive indices it will only work for light rays that hit AB near the bottom.

Section 2.7: Photons

Q1 (a) A photon is a particle of light. [It has energy hf where h is the Planck constant.]

(b) Photon energy = hf.
If there are n photons per second, energy transfer per second = nhf.

Q2 A line emission spectrum consists of a series of narrow bands of light, each with a particular frequency / colour / wavelength. It is produced by hot, low-pressure, gases which consist of individual atoms and/or molecules. It can be displayed by passing the light through a diffraction grating or a prism and projecting the resulting beams onto a screen.

Q3 Light from the surface [photosphere] passes through a tenuous atmosphere of gas. Gas atoms absorb photons of light at specific wavelengths, corresponding to the difference in energy levels of the atoms. The excited atoms emit photons of the same wavelength but in random directions, so that if the spectrum of the light is observed from the Earth it appears as the continuous spectrum of light from the photosphere with a series of dark lines across it – the absorption spectrum.

Q4 (a) De Broglie wavelength, $\lambda = \frac{h}{p}$. The momentum, $p = \sqrt{2mE} = \sqrt{2meV}$

So $\lambda = \frac{h}{\sqrt{2meV}} = \frac{6.63 \times 10^{-34}}{\sqrt{2 \times 9.11 \times 10^{-31} \times 1.60 \times 10^{-19} \times 2400}} = 2.51 \times 10^{-11}$ m

(b) From the calculation in part (a) we see that as the voltage is increased, the wavelength of the electrons decreases. Hence the diffraction pattern gets smaller [the sine of the angle to the centre of the diffracted beam is proportional to the wavelength].

Q5 (a) (i) The photon energy is the difference in the two energy levels.
So energy = 12.1 eV – 10.2 eV = 1.9 eV

(ii) $E = hf = h\dfrac{c}{\lambda}$

So $\lambda = \dfrac{hc}{E} = \dfrac{6.63 \times 10^{-34} \times 3.00 \times 10^8}{1.9 \times 1.60 \times 10^{-19}} = 6.5 \times 10^{-7}$ m; visible (red)

(b) (i) Greatest energy = 12.75 – 0.0 = 12.75 eV
$$= 12.75 \times 1.60 \times 10^{-19} \text{ J}$$
$$= 2.0 \times 10^{-18} \text{ J}$$

(ii) Ultra-violet.

Q6 (a) Einstein's photoelectric equation can be written $E_{k\,max} = hf - \phi$. It is an expression of conservation of energy: the maximum electron energy $E_{k\,max}$ is equal to the photon energy (hf) minus the minimum energy needed to remove an electron from the surface (the work function, ϕ). It provided a firm experimental foundation for the wave–particle duality of quantum theory by showing that light has properties that are only explicable if it consists of a stream of particles with energy proportional to the frequency (a wave property).

To obtain the results, the photocell is illuminated with monochromatic radiation (of high enough frequency) and the supply pd increased until the photo-current just becomes zero. This pd value is multiplied by e (the electronic charge) to give $E_{k\,max}$. This is repeated for several frequencies and the graph drawn.

(b) All photons of a monochromatic beam have the same energy. The intensity of the beam is proportional to the number of photons per second passing a point [incident on the emitting surface]. Each photon has the same probability of causing the emission of an electron, so the number of electrons emitted per second and received on the collecting electrode is proportional to the intensity. Hence the current is also proportional.

Q7 (a) The photons in the beam each have a momentum, p, given by the de Broglie relationship, $p = h/\lambda$. As each is reflected from the mirror, it suffers a change in momentum. The space probe receives an equal and opposite change of momentum, i.e. a force.

(b) (i) $p = \dfrac{h}{\lambda}$. Also $E = hf = \dfrac{hc}{\lambda}$.

Substituting for $\lambda \longrightarrow p = h \times \dfrac{1}{\lambda} = h \times \dfrac{E}{hc}$, so $E = pc$.

(ii) Change in momentum per second of each photon = $2 \times \dfrac{E}{c}$ because of the reflection.

So force on probe = $2\dfrac{E}{c} \times$ number of photons per second = $2\dfrac{P}{c} = 2 \times \dfrac{15\,000}{3.00 \times 10^8}$ N
$$= 0.10 \text{ mN}$$

So $a = \dfrac{F}{m} = \dfrac{0.1 \times 10^{-3}}{2.3} = 4.3 \times 10^{-5}$ m s^{-2}.

(iii) (I) $v = u + at$, so speed = $4.3 \times 10^{-5} \times (86\,400 \times 365) = 1400$ m s^{-1}

(II) $x = ut + \dfrac{1}{2}at^2 = 0.5 \times 4.3 \times 10^{-5} \times (86\,400 \times 365)^2$
$$= 2.1 \times 10^{10} \text{ m [21 million km]}$$

(iv) The laser is aimed at the mirror which reflects the beam back towards the probe, thus producing a reverse thrust.

(c) (i) Assume that the diameter of the window is 2 mm.

Then $\theta = \dfrac{2\lambda}{d} = \dfrac{2 \times 500 \times 10^{-9} \text{ m}}{2 \times 10^{-3} \text{m}} = 5 \times 10^{-4}$ rad.

So, from the back of the lecture room to the front, the diameter, D, of the circle of the spot is given by:
$D = 5 \times 10^{-4} \times 10$ m = 5×10^{-3} m = 5 mm.
A spot of light of diameter 5 mm is small enough to be useful as a pointer, so diffraction is not a problem.

(ii) [There are many ways of answering this AO3 question. This is one of them.]

If the power incident on the spacecraft is 10% of the original, then the area of the beam is 10 × the area of the receiving surface, so the diameter is $\sqrt{10} \times 100$ m = 320 m.

The diffraction angle $\theta = \dfrac{2\lambda}{d} = \dfrac{2 \times 500 \times 10^{-9}\text{ m}}{1\text{ m}} = 1 \times 10^{-6}$ rad.

So distance for the diameter of beam to be 320 m = $\dfrac{320\text{ m}}{1 \times 10^{-6}\text{rad}} = 3.2 \times 10^{8}$ m

This distance, 300 000 km is much less than the distance travelled in 1 year calculated in part (b), so the speed achieved will also be much less and the spacecraft will take many years to reach Mars.

Section 2.8: Lasers

Q1 An atom is in an excited state. A photon with energy equal to the energy difference between the excited state and an unoccupied lower energy state triggers the atom to change to a lower energy state. This releases a second photon – *stimulated emission*.

Q2 (a) If there are two energy states in a system (e.g. a collection of atoms) normally the lower energy state has a greater population. A population inversion is when there are more atoms in the higher energy state than in the lower.

(b) An incident photon (of the correct energy) can be absorbed and cause an atom in the lower energy state to jump to the higher. An identical photon can cause an atom in the higher energy state to fall to the lower. In a population with equal numbers in the lower and higher energy states, the optical pumping will cause equal numbers of transitions in the two directions, so the number of atoms in the higher energy state cannot increase above this level.

Q3

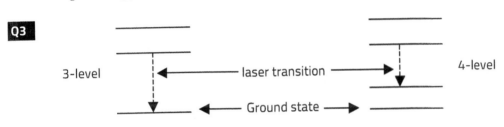

In a 3-level system the lower level in the laser transition is the ground state. So, in order to get a population inversion with an excited state, at least half the atoms must be in the upper level. In a 4-level system, the lower level in the laser transition is a normally empty excited level, so only a small proportion of atoms need to be put into the upper level, which is easier to achieve.

Q4 The pumping level needs to empty quickly so:

(i) the population in the upper level builds up quickly, and

(ii) to reduce stimulated emission between the pumping and ground levels [OR, equivalently, to stop the top level filling up (making pumping less efficient)].

Q5 Uses: in optical fibres and DVD reading heads
Advantages: small, efficient (low power use) [also very cheap]

Q6 (a) Laser transition $E_3 \longrightarrow E_2$,

so energy = 19.5 eV – 17.4 eV = 2.1 eV = 2.1 × 1.60 × 10^{-19} = 3.36 × 10^{-19} J

$f = \dfrac{E}{h} = \dfrac{3.36 \times 10^{-19}\text{ J}}{6.63 \times 10^{-34}\text{ Js}} = 5.1 \times 10^{14}$ Hz (2 sf)

(b) The transition is stimulated by a photon of energy 2.1 eV. The system is chosen so that level E3 is a metastable energy level, i.e. it has a long lifetime, so the probability of a spontaneous emission is low. However, a photon of energy 2.1 eV has a high probability of triggering emission. [Also, there is so much light inside a laser that the photon will be stimulated long before spontaneous emission is likely to occur.]

Unit 2 Answers

(c) To produce each laser photon of energy 2.1 eV, the pumping energy required is 20.5 eV. Some of the pumped energy will be wasted by a transition back down to the ground state.

Hence the maximum efficiency = $\dfrac{2.1\,\text{eV}}{20.5\,\text{eV}} \times 100\% = 10.2\%$. So Joel is correct to 2 sf [and some energy will inevitably be lost anyway].

(d) In order to establish a population inversion between E_2 and E_3, the population of E_2 must be kept as low as possible. Hence the second energy level must have a low lifetime, so that it can empty back down to the ground state as quickly as possible and Nigella is wrong.

Q7 (a) Pumping energy = $4.8 \times 10^{-19}\,\text{J} = \dfrac{4.8 \times 10^{-19}}{1.60 \times 10^{-19}}\,\text{eV} = 3.0\,\text{eV}$

Violet [but a marking scheme would accept blue, indigo or violet].

(b) E_3 the pumped level should have a short lifetime and E_2 a much longer one. This is to allow a population inversion to be established between E_2 and E_1.

(c) $E = hf$ and $c = f\lambda$ ∴ $\lambda = \dfrac{hc}{E} = \dfrac{6.63 \times 10^{-34} \times 3.00 \times 10^8}{3.1 \times 10^{-19}} = 6.4 \times 10^{-7}\,\text{m}$

(d) Same frequency [or wavelength or energy]
Same phase
Same direction
Same polarisation [NB only 3 asked for – don't give 4 in case one is wrong!]

(e) Paula is correct. To achieve a population inversion between E_1 and E_2, the population of E_2 (i.e. N_2) must be greater than N_1. This happens most efficiently if all the pumped atoms immediately drop down to E_2. N_1 is more than 50% of the total, then N_2 must be less and a population inversion doesn't exist.

Q8 (a) The exiting laser beam removes some photons from the cavity, so to keep the number as high as possible, there should be no losses at the left hand mirror.

(b) Some photons need to escape to form the exiting laser beam.

(c) (i) More photons (90%) are reflected on the left-hand mirror than on the right (40%), so there is more momentum change of the photons per second on the left than the right. Hence, by Newton's second law, there is a greater force on the left.

(c) (ii) 60% of photons incident on right-hand mirror escape.

∴ Power of photons incident on mirror = $\dfrac{5.0\,\text{mW}}{0.6} = 8.33\,\text{mW}$

∴ Power reflected = 8.33 mW – 5.0 mW = 2.33 mW

∴ Force exerted = $2 \times \dfrac{P}{c} = 1.55 \times 10^{-11}\,\text{N}$

∴ Pressure = $\dfrac{F}{A} = \dfrac{1.55 \times 10^{-11}\,\text{N}}{0.94 \times 10^{-6}\,\text{m}^2} = 1.7 \times 10^{-5}\,\text{Pa}$ (2 sf)

(d) Consider N photons being reflected from right mirror. Assume number hitting left-hand mirror = kN, where k is the amplification factor for one pass. The number reflected from left-hand mirror = kN. Number hitting right-hand mirror = k^2N. Then number reflected from right-hand mirror = $0.95 \times k^2N$.

∴ To achieve equilibrium $0.95k^2 N = N$

∴ $k = \dfrac{1}{\sqrt{0.95}} = 1.026$, so the increase is 2.6% and Helena is correct – she's a bright cookie!

Physics 1 practice paper

Q1 (a)

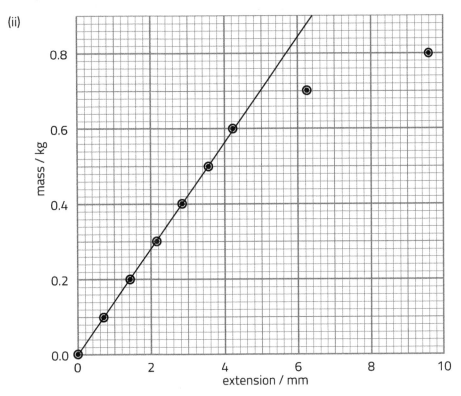

If equal and opposite forces, F, are applied to the rod, as shown, the tensile stress is defined as F/A where A is the cross-sectional area of the rod.

(b) (i) Aled used either a digital calliper or micrometer. This is because the diameter is measured to 0.01 mm.

(ii)

For the straight-line portion of the graph, when mass = 0.700 kg, the extension is 5.00 mm

At this point: stress $= \dfrac{mg}{csa} = \dfrac{0.700 \times 9.81}{\pi \times (0.10 \times 10^{-3})^2} = 2.186 \times 10^8$ Pa

and strain $= \dfrac{5.00 \times 10^{-3}}{2.500} = 2.00 \times 10^{-3}$

$\therefore E = \dfrac{2.186 \times 10^8}{2.00 \times 10^{-3}} = 1.093 \times 10^{11}$ Pa

Only 2 sf justified because of diameter reading, $\therefore 1.1 \times 10^{11}$ Pa.

(iii) (I) Remove the load and measure the length of the wire. If longer than originally, inelastic stretch had occurred.

(II) Bonds are broken at the site of an edge dislocation (an extra half-plane). The stress causes movement of the planes which does not reverse when the stress is removed.

(b) This knowledge has allowed engineers to develop novel materials which are able to withstand high stresses without plastic deformation. An example is the development of single-crystal, high-purity materials for turbine blades, which can operate at high revs without large deformation. The absence of foreign atoms avoids the production of dislocations at moderate stresses.

Q2 (a) (i) Moment = force × distance = (44 × 9.81) N × 1.2 m = 518 N m clockwise.

(ii) If tension = T, the anticlockwise moment = $T \times 2.4 \sin 35°$
∴ For equilibrium $2.4T \sin 35° = 518$
∴ $T = 376$ N ~ 400 N.

(iii) Horizontal component of hinge force is equal and opposite to the horizontal component of the tension force, for equilibrium.

$$\therefore F_{horiz} = 376 \cos 35° = 308 \text{ N (to the right)}$$

(b) The maximum clockwise moment is when the student is at the right-hand end of the bridge. At this point: Total clockwise moment = $518 + 68 \times 9.81 \times 2.4 = 2119$ N m.

$$\text{Required tension} = \frac{2119}{2.4 \sin 35°} = 1540.$$

This is greater than 1500 N so it is unsafe.

Q3 (a) (i) Force is provided by the gravitational force on the 0.20 kg = $0.20 \times 9.81 = 1.962$ N. The accelerating mass = $1.0 + 0.20 = 1.20$ kg

$$\therefore \text{acceleration} = \frac{1.962}{1.20} = 1.64 \text{ m s}^{-2} \sim 1.6 \text{ m s}^{-2}.$$

(ii) The 0.20 kg mass is accelerating downwards, so there must be a resultant force downwards. Hence the upward force due to the tension must be less than the downward force due to gravity.

(b) Displacement in 0.150 s = 11.8 − 10.0 = 1.8 cm.
Displacement in 0.300 s = 17.3 − 10.0 = 7.3 cm.
For constant acceleration from rest, $x \propto t^2$, so double time $\longrightarrow 4 \times$ displacement
7.3 ~ 4 × 1.8, so constant acceleration.
$$x = \frac{1}{2}at^2, \text{ so } a = \frac{2x}{t^2} = \frac{14.6 \text{ cm}}{0.3^2} = 162 \text{ cm s}^{-2} \sim 1.64 \text{ m s}^{-2} \text{ so consistent.}$$

Q4 (a) The algebraic sum of the momenta of the bodies in an isolated system [i.e. with no external forces acting] is constant.

(b) (i) Assuming no friction:
Momentum acquired by block = $0.42 \times (28.0 - (-9.0)) = 15.54$ N s
$$\therefore E_k = \frac{p^2}{2m} = \frac{(15.54)^2}{10} = 24.1 \text{ J}$$
[Alternative to the last line: $v = \frac{15.54}{5.0} = 3.108$ m s^{-1} so KE = $\frac{1}{2} \times 5.0 \times (3.108)^2 = 24.1$ J]

(ii) Initial KE = $\frac{1}{2} \times 0.42 \times (28.0)^2 = 164$ J
KE of 0.42 kg ball after collision = $\frac{1}{2} \times 0.42 \times (9.0)^2 = 17.0$ J
∴ Total KE after collision = 24.1 + 17.0 = 41.1 J
∴ KE lost, so the collision is inelastic.

(iii) Frictional force = $\frac{24.1 \text{ J}}{3.6 \text{ m}} = 6.7$ N

Q5 As each ball accelerates downwards because of the force of gravity, it experiences an upward force of air resistance, which increases with velocity. As it reaches a speed at which the two forces are equal in magnitude, the resultant force becomes zero and no further acceleration occurs – this is the so-called terminal velocity.

The heavier ball needs a greater air resistance force to balance the greater gravitational force (its weight). Therefore, because the shape and surface area are the same for the two balls, the balance does not happen until the speed is greater.

Q6 (a) (i) (I) Speed = $\frac{\text{semi-circumference}}{\text{time}} = \frac{\pi \times 60 \text{ m}}{15 \text{ s}} = 12.6 \text{ m s}^{-1}$

(II) Mean velocity = $\frac{\text{displacement}}{\text{time}} = \frac{120 \text{ m}}{15 \text{ s}}$ East = 8.0 m s^{-1} E

(ii) At A the velocity is 12.6 m s^{-1} N. At B the velocity is 12.6 m s^{-1} S, i.e. −12.6 m s^{-1} N. There is a change in velocity of 25.2 m s^{-1} S, hence an acceleration.

(b) (i) velocity upwards / m s^{-1}

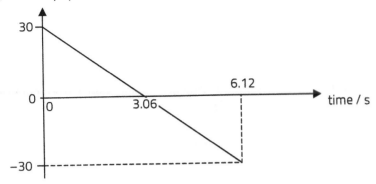

(ii) Distance = $2 \times (\frac{1}{2} \times 3.06 \times 30) = 91.8$ m.

Q7 (a) Peak wavelength = 418 nm

So by Wien's law, $T = \dfrac{2.90 \times 10^{-3}\,\text{m K}}{418 \times 10^{-9}\,\text{m}} = 6940$ K, i.e. 7 000 K approx.

(b) Luminosity $\propto$ surface area $\times$ temperature^4

$$\frac{L_{Can}}{L_{Sol}} = \left(\frac{r_{Can}}{r_{Sol}}\right)^2 \left(\frac{T_{Can}}{T_{Sol}}\right)^4$$

$$\frac{r_{Can}}{r_{Sol}} = \sqrt{\frac{L_{Can}}{L_{Sol}}} \times \left(\frac{T_{Sol}}{T_{Can}}\right)^2 = \sqrt{10700} \times \left(\frac{5780}{6940}\right)^2 = 72$$

(c) Its radius is much larger than that of the Sun, so it is a giant. However, the peak wavelength is about 400 nm, which is at the blue end of the visible spectrum, so its colour will be bluish-white. Hence it is a blue giant and Megan is partly correct.

Q8 (a) (i) A particle comprising 3 quarks.

(ii) A particle comprising a quark and an anti-quark.

(b) (i) Baryon number (B) is conserved.
Protons have a baryon number of 1 and pions a baryon number of 0. To conserve B, the baryon number of x must be 1, so it is a baryon.
Charge is conserved.
The protons and the pion each have a charge of +1, so the charge of x is 0.
As x is a first-generation particle it could be a neutron or a Δ^0.

(ii) u quarks. LHS $U = 2 + 2 = 4$; RHS $U = 2 + 1 + 1 = 4$, $\therefore$ conserved.
d quarks: LHS $D = 1 + 1 = 2$; RHS $D = 1 + 2 + (-1) = 2$, $\therefore$ conserved.

(iii) Conservation of quark numbers means that either strong or electromagnetic is possible. Strong is much more likely than e-m as it is 'stronger'.

(c) (i) It has a baryon number of 0 (for B to be conserved), a charge of 0 (for Q to be conserved) and a lepton number of 1 (for L to be conserved, because e$^+$ has $L = -1$).

(ii) 2 reasons: First, the interaction only happens 'occasionally'. Secondly (and definitively), quark numbers are not separately conserved. The proton has quark structure uud; the deuteron has (uud) + (udd). Hence U decreases by 1, D increases by 1.

Physics 2 practice paper

Q1 (a) Electric current is the rate of flow of electric charge.

(b) Conduction of electricity in a metal consists of the passage of 'free' electrons through the metal. Free electrons, although coming from the atoms of the metal, are no longer attached to individual atoms. The free electrons are in constant translational motion, sharing the random energy of the vibrating ions with which they are constantly colliding. When a pd is placed across a piece of metal the free electrons experience forces towards the positive terminal of the pd. The collisions prevent the electrons from continuously accelerating; instead a mean 'drift' velocity in that direction is superimposed on their random thermal motion.

An increase in temperature implies an increase in random energy, so collisions between electrons and ions are more frequent. Therefore, the mean drift velocity of the free electrons decreases for a given applied pd. Therefore the current decreases, and the metal's resistance increases.

(c) (i) $R = \dfrac{\rho l}{A} = \dfrac{5.6 \times 10^{-8}\ \Omega\,\text{m} \times 0.65\,\text{m}}{2.75 \times 10^{-5}\ \text{m}^2} = 1.32\ \text{m}\Omega\ (1.32 \times 10^{-3}\ \Omega)$

(ii) $I = \dfrac{V}{R} = \dfrac{1.65\,\text{V}}{0.00132\,\Omega} = 1\,247\ \text{A}$

$v = \dfrac{I}{nAe} = \dfrac{1247\,\text{A}}{8.25 \times 10^{28}\ \text{m}^{-3} \times 2.75 \times 10^{-5}\ \text{m}^2 \times 1.60 \times 10^{-19}\ \text{C}}$

$= 3.43\ \text{mm s}^{-1}$

Q2 (a)

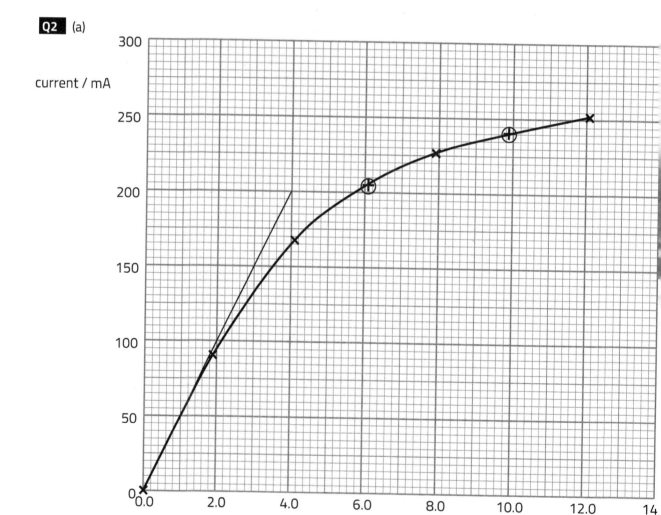

(b) (i) $R = \dfrac{1}{\text{gradient at origin}} = \dfrac{4.00\,\text{V}}{0.200\,\text{A}} = 20\ \Omega$

(ii) $R = \dfrac{12.0\,\text{V}}{0.252\,\text{A}} = 48\ \Omega$

(c) (i)   The results and the graph show resistance increasing with pd. This is to be expected because the filament's temperature increases with increasing pd, resulting in more frequent collisions between ions and free electrons. The temperature is such that the filament emits radiation at the same rate as electrical power is dissipated; so a greater pd causes greater current, greater power dissipation, a higher temperature and greater resistance!

(ii)   A particular pd will always give the same current. Adjusting the potential divider will give a different (but perfectly valid) pair of readings. Attempting to come back to a previous pair will be tricky to do and merely repetitive if achieved.

(d) (i)   Current of any chosen value, once started, will continue even with no pd present. That renders meaningless the idea of current plotted against simultaneous pd.

(ii)   (e.g.) The conductors in the electromagnets in the Large Hadron Collider.

Q3 (a) E is the chemical energy transferred in the cell per unit charge.
IR is the energy transferred to the external resistance per unit charge.
Ir is the energy transferred to the internal resistance per unit charge.
Energy is conserved, so $EI = I^2R + I^2r$ and $EI = IR + Ir$

(b) (i) $I = \dfrac{E}{R+r} = \dfrac{1.65\ \text{V}}{0.20\ \Omega + 0.20\ \Omega} = 4.1\ \text{A}$

(ii) $P = I^2R = (4.13\ \text{A})^2 \times 0.20\ \Omega = 3.41\ \text{W}$

(c) [Example of a possible approach]

Try $R = 0.15\ \Omega$. Then $P = \left(\dfrac{1.65\ \text{V}}{0.15\ \Omega + 0.20\ \Omega}\right)^2 \times 0.15\ \Omega = 3.33\ \text{W}$

Try $R = 0.25\ \Omega$. Then $P = \left(\dfrac{1.65\ \text{V}}{0.25\ \Omega + 0.20\ \Omega}\right)^2 \times 0.25\ \Omega = 3.36\ \text{W}$

Both these values of power are smaller than for $R = 0.20\ \Omega$, so it seems likely that the maximum value of power is at $R = 0.20\ \Omega$.

Q4 (a) In a transverse wave, the oscillations are at right angles to the direction of propagation; in a longitudinal wave, the vibrations are parallel to the direction of propagation.

(b) (i)   0.20 s

(ii)   5.0 Hz

(c) (i) $\lambda = (3.15 - 0.95)\ \text{cm} = 2.20\ \text{cm}$
$v = f\lambda = 5.0\ \text{Hz} \times 2.20\ \text{cm} = 11\ \text{cm s}^{-1}$

(ii)

displacement / cm

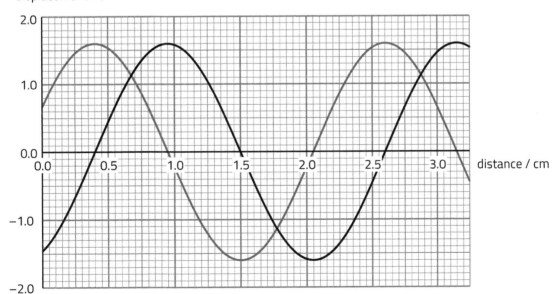

Q5 (a) Strongly suggest that light has wave properties [is a wave]

(b) Fringe separation = $\dfrac{65 \text{ mm}}{9}$ = 7.2 mm

$\lambda = \dfrac{0.25 \text{ mm} \times 7.2 \text{ mm}}{2.250 \text{ m}}$ = 800 nm / 0.80 µm (2 sf)

(c) Light from one slit has to travel one wavelength further to reach A than light from the other slit. This means that light from the two slits arrives at A peak-on-peak and trough-on-trough, so there is constructive interference and, according to the principle of superposition, twice the wave amplitude that there would be due to one slit alone.

(d)

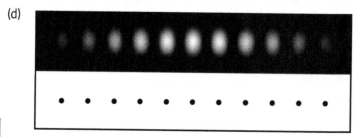

Q6

(a) Snell's law: $n \sin \theta$ = constant $\therefore 1.00 \sin 22.2° = 1.64 \sin \theta$

$\therefore \theta = \sin^{-1}\left[\dfrac{\sin 22.2°}{1.64}\right]$ = 13.3°

(b) Angle of incidence at core-cladding boundary ϕ = 90.0° − 13.3° = 76.7°
But critical angle C is given by 1.64 sin C = 1.61 sin 90°
So $C = \sin^{-1}\left(\dfrac{1.61 \times 1}{1.64}\right)$ = 79.0°
So $\phi < C$ so there will only be partial reflection and so there will be much less light in the core after a few reflections.

(c) [There is no definitive 'right' answer; many points could be made, e.g....] Considering only the carbon dioxide emission, which contributes to anthropogenic global warming, it depends upon how much energy is needed in the purification process for the sand (the main raw material for the silica out of which the glass fibres are made) and for the production of the fibres of both types. Assuming that the use of the products of oil refining for optical fibres does not add significantly to oil extraction and burning (because it is only a by-product) oil-based optical fibres are do not significantly affect the environment because of greenhouse emissions. Glass optical fibres also use plastics for the sheaths around the fibres so the disposal of these could be a problem with the release of plastics into the environment. There is probably little distinction. [You could get full marks for a much shorter answer.]

Q7 (a) 3-level system: Laser transition = 1.79 eV.
$E = \dfrac{hc}{\lambda} \therefore \lambda = \dfrac{6.63 \times 10^{-34} \text{ Js} \times 3.00 \times 10^{8} \text{ m s}^{-1}}{1.79 \times 1.60 \times 10^{-19} \text{ J}}$ = 6.94 × 10⁻⁷ m

4 level system: Laser transition = 1.42 − 0.25 = 1.17 eV

$\therefore \lambda = \dfrac{6.63 \times 10^{-34} \text{ Js} \times 3.00 \times 10^{8} \text{ m s}^{-1}}{1.17 \times 1.60 \times 10^{-19} \text{ J}}$ = 1.06 × 10⁻⁶ m

(b) In a 3-level system, the bottom level of the laser transition is the ground state. Hence to achieve population inversion more than 50% of the atoms must be placed in the upper state. In the 4-level system, the bottom level of the laser transition is an excited level, which has an initial population of zero. Hence putting any atoms into the top level of the laser transition produces a population inversion, so the energy needed to start the laser is much less.

(c) Photons of energy corresponding to the lasing transition cause stimulated emission from the amplifying medium: the emission of photons of the same energy, phase and direction of travel. Those travelling parallel to the tube axis are reflected back and forth by the mirrors at either end, causing more stimulated emission. Thus the beam builds in intensity until a maximum, limited by the rate of pumping and by escape of 1% of the photons incident on the 99% reflecting mirror. The escaping photons form the laser beam.

Q8 (a) The photon energy is hf. The maximum kinetic energy, $E_{k\,max}$, of the photoelectrons, is equal to this energy minus the minimum energy needed to remove an electron from the metal surface – the work function, ϕ.

(b) (i) $\phi =$ – the intercept on the $E_{k\,max}$ axis $= 3.4 \times 10^{-19}$ J

$h =$ gradient of graph $= \dfrac{(3.3-(-3.4)) \times 10^{-19}\text{J}}{10 \times 10^{14}\text{ Hz}} = 6.7 \times 10^{-34}$ J s

(ii) There is no noticeable scatter about the line of best fit.
The line of best fit is straight, its slope gives a value of h in close agreement with that found by other methods and its intercept is negative. So it accords with Einstein's equation.